Lecture Notes in Artificial Intelligence   1011

Subseries of Lecture Notes in Computer Science
Edited by J. G. Carbonell and J. Siekmann

Lecture Notes in Computer Science

Edited by G. Goos, J. Hartmanis and J. van Leeuwen

Springer
Berlin
Heidelberg
New York
Barcelona
Budapest
Hong Kong
London
Milan
Paris
Santa Clara
Singapore
Tokyo

Takeshi Furuhashi (Ed.)

Advances in Fuzzy Logic, Neural Networks and Genetic Algorithms

IEEE/Nagoya-University
World Wisepersons Workshop
Nagoya, Japan, August 9-10, 1994
Selected Papers

 Springer

Series Editors

Jaime G. Carbonell, Carnegie Mellon University, USA

Jörg Siekmann, University of Saarland, DFKI, Germany

Volume Editor

Takeshi Furuhashi
Department of Information Electronics
Nagoya University, School of Engineering
Furo-cho, Chikusa-ku, Nagoya, 464-01 Japan

Cataloging-in-Publication Data applied for

Die Deutsche Bibliothek - CIP-Einheitsaufnahme

Advances in fuzzy logic, neural networks and genetic algorithms
: selected papers / IEEE/Nagoya-University World Wisepersons
Workshop, Nagoya, Japan, August 1994. Takeshi Furuhashi
(ed.). - Berlin ; Heidelberg ; New York ; Barcelona ; Budapest ;
Hong Kong ; London ; Milan ; Paris ; Tokyo : Springer, 1995
  (Lecture notes in computer science ; Vol. 1011 : Lecture notes in
  artificial intelligence)
  ISBN 3-540-60607-6
NE: Furuhashi, Takeshi [Hrsg.]; World Wisepersons Workshop <3, 1994,
  Nagoya>; Industrial Electronics Society; GT

CR Subject Classification (1991): I.2, F.1-2

ISBN 3-540-60607-6 Springer-Verlag Berlin Heidelberg New York

Typesetting: Camera ready by author
SPIN 10487204   06/3142 – 5 4 3 2 1 0   Printed on acid-free paper

Preface

Fuzzy logic is one of the key technologies for representing human knowledge in the brain and for constructing adaptive systems. The difficulties of fuzzy logic are knowledge acquisition from experts and/or knowledge finding for unknown tasks. Neural networks and genetic algorithms are also gaining attention for their potential for efficient knowledge acquisition/finding. Combining the technologies of fuzzy logic and neural networks/genetic algorithms is expected to open a new paradigm of machine learning for realization of human-like information processing systems. This is the background against which the 1994 IEEE/Nagoya University World Wisepersons Workshop on Fuzzy Logic and Neural Networks/Genetic Algorithms(WWW'94) was organized.

The WWW'94 was the third workshop in the WWW series. The first workshop of this series was opened by Prof. Fukuda (Nagoya University) with the support and cooperation of IEEE/Industrial Electronics Society. Addressing the theme of "Multiple Distributed Robotic Systems" (Chairman: Assoc. Prof. Kosuge, Dept. of Mechano-Informatics and Systems, Nagoya University), it was held in July, 1993, and attracted many excellent participants and papers. The second in the series (Chairman: Assistant Prof. Arai, Dept. of Mechano-Informatics and Systems, Nagoya University) was held in October, 1993, with the theme of "Learning and Adaptative Systems".

The third World Wisepersons Workshop was held in August, 1994, to give attendees a place to have a lively discussion on the subject of "Present and Future Situation of the Combination Technologies of Fuzzy Logic and Neural Networks/Genetic Algorithm", an area which has made remarkable progress recently. The steering committee of WWW'94 has accepted 19 excellent papers. More than 50 papers were submitted to the steering committee and we had to make severe decisions to reject many excellent papers.

As the editor for the Selected Papers of WWW'94, I selected fourteen excellent papers which were revised, extended, and rigorously reviewed by two reviewers each. The first six papers are related to combination technologies of fuzzy logic and neural networks. The next two papers discuss interesting applications of fuzzy systems. The ninth to twelfth papers describe challenging works for combining fuzzy logics and genetic algorithms. The last two papers present significant applications of fuzzy-GA. I am now proud to have these papers published in the LNCS/LNAI series.

Nagoya, August 1995   Takeshi Furuhashi

Contents

3. Fuzzy - Genetic Algorithms

4. Fuzzy - GA Applications

Fuzzy associative memory system and its application to multi-modal interface

Hirohide Ushida [+], Tomohiko Sato [+],
Toru Yamaguchi [++] and Tomohiro Takagi [+++]

[+] Laboratory for International Fuzzy Engineering Research
[++] Faculty of Engineering, Utsunomiya University
[+++] Multimedia Software Development Office, Matsushita Electric Industrial Co., LTD

Abstract A construction method and a refinement method of fuzzy knowledge are proposed in order to apply them to intelligent multi-modal interfaces. This paper supposes that the interface requires the following three functions at least. 1) A function that constructs knowledge using instances instead of if-then rules. 2) A function that transforms mutually between the upper conceptual label represented by words and the lower conceptual label represented by physical values in order to realize multi-modality. 3) A function that refines the knowledge using macro qualitative instruction such as a human learning process. This paper proposes the methods using Fuzzy Associative Memory Organizing Units System (FAMOUS) in order to realize these functions and applies them to estimation of human movements. Experimental results show that the proposed methods provide the three functions and are suitable for application to intelligent multi-modal interfaces.

1 Introduction

In conventional human-machine interfaces, a person gives a command and numerical values to a machine and accepts numerical values from the machine. The commands and the values are represented in micro orders. Using this method to process very complex or "intelligent" tasks would require vast amount of time and personal resources. In contrast, humans use multi-modal macro qualitative representation to communicate. Multi-modality refers to the use of body language and facial expression as well as linguistic language. People communicate efficiently and smoothly by means of this multi-modality. Interfaces which understand and generate human macro qualitative representation need to be able to facilitate this kind of communication between people and machines. This paper supposes that the interface requires the following three functions at least. 1) A function that constructs knowledge using instances instead of if-then rules. 2) A function that transforms mutually between the upper conceptual labels represented by words and the lower conceptual labels represented by physical values in order to realize multi-modality. 3) A function that refines the knowledge using macro qualitative instruction such as a human learning process.

The macro qualitative representation includes fuzziness. Fuzzy inference [1] is a method to process the fuzziness which can transform micro numerical values into macro linguistic expression. But the conventional fuzzy inference does not have the above-mentioned function 1) because the knowledge of the fuzzy inference is described by if-then rules. The feed forward neural networks which have been developed [2] have the function 1) because the knowledge is captured using numerical values of instances experimentally. But it is difficult for these neural networks to have the functions 2) and 3). Furthermore, the knowledge representation of the neural networks is not clear, and such clearness is required for the interface be able to analyze the process.

To solve these problems methods which use Fuzzy Associative Memory Organizing Units System (FAMOUS) [3] are described in this paper. The FAMOUS has a hierarchical knowledge structure using bidirectional associative memories (BAMs) [4] and its knowledge representation is clear. The FAMOUS has three features as follows. i) The knowledge is constructed using instances. ii) It transforms mutually between the upper conceptual labels and the lower conceptual labels. iii) It is available for knowledge refinement by means of macro qualitative instructions. These features enable the system to perform functions 1), 2) and 3) described above. This paper shows a knowledge construction method and a knowledge refinement method which are applied to estimate human movement.

Section 2 of this paper describes the construction method and the inference method of the FAMOUS. Section 3 describes the FAMOUS-based model which recognizes and generates human facial expressions using feature i), ii), and iii). Section 4 describes the knowledge construction method to be applied to estimate human physical performance using feature i). Section 5 describes the knowledge refinement method using features ii) and iii). The refinement process is called "conceptual level learning". In Section 6, the methods are applied to a system which estimates human tennis shots, and experimental results are shown in Section 7.

2 Fuzzy associative memory organizing units system

Fuzzy Associative Memory Organizing Units System (FAMOUS) [3] is a kind of set of associative memories. It consists of several bidirectional associative memories (BAMs) [4]. Fuzzy associative inference is driven by node activation propagation in the BAMs. BAM has two layers La and Lb which are connected to one another (Fig. 1). The strength of the connection is represented by the correlation matrix M. The BAM recall process is termed reverberation and is carried out according to Eq. (1).

$$Y(t) = S(MX(t)), \quad X(t+1) = S(M^T Y(t)) \tag{1}$$

In Eq. (1), $X(t) = [x_1(t), x_2(t), ..., x_m(t)]^T$ and $Y(t) = [y_1(t), y_2(t), ..., y_n(t)]^T$. The $x_i(t)$ is an activation of node a_i and the $y_i(t)$ is an activation of node b_i at reverberation step t. The $X(t)$ and $Y(t)$ are activation vectors on the layers. $S(\cdot)$ is the sigmoid function of each node. The correlation matrices M and M^T are given by Eq. (2):

$$M = \sum_{k=1}^{p} \beta_k Y_k X_k^T, \quad M^T = \sum_{k=1}^{p} \beta_k X_k Y_k^T \tag{2}$$

In Eq. (2), X_k, Y_k (k =1 to p) are stored pairs of vectors in a BAM. β_i is an association parameter, in which each element of X_k, Y_k [0, 1] is usually converted to a bipolar element [-1, 1] based on the BAM energy function. The BAM recalls the

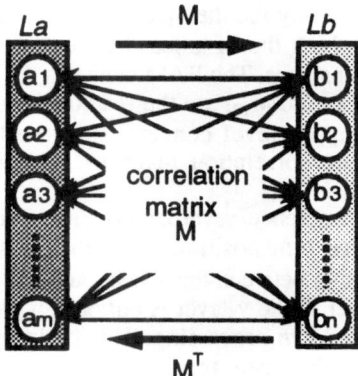

Fig. 1 Bidirectional Associative Memory.

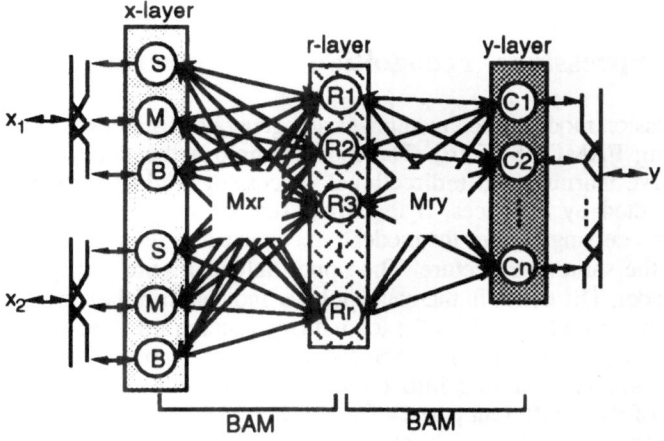

Fig. 2 Configuration of FAMOUS.

pair which is the most suitable for the input condition as a result of decreasing fuzzy entropy, which describes the degree of fuzziness of the inference output.

The memorizing method based on Eq. (2) is a kind of inductive learning. This method is, however, limited in terms of the pairs that are memorized. That is, the pairs to be memorized cannot be similar. If there is a similarity between pairs or, if complex patterns have to be memorized, then several BAMs are used to construct a FAMOUS.

There are two types of FAMOUS. One is rule-based FAMOUS and the other is instance-based FAMOUS. The first one is constructed using fuzzy rules and the second one is constructed using instances. Figure 2 shows an example of a three layered FAMOUS consisting of an x-layer, r-layer, and y-layer. Each node on the x-layer represents the fuzzy label given by the if-part of the fuzzy rules or the fuzzy label of the lower level feature obtained from the instances. Each node on the y-layer

4

represents the fuzzy label given by the then-part of the fuzzy rules or the fuzzy label of the upper level concept which the instances belong to. Each node on the r-layer represents a fuzzy rule or an instance. The BAM connects the x-layer to the r-layer and the r-layer to the y-layer. M_{xr} and M_{ry} are the resulting correlation matrices. Thus, the relations between the upper level concepts and the lower level values are constructed by means of the correlation matrices. This is the feature i) of the FAMOUS.

During fuzzy associative inference, reverberation is performed with the correlation matrices M_{xr}, M_{ry}, and each transposition correlation matrix. This reverberation permits bidirectional processing between the x-layer and the y-layer. The propagation of activations from the x-layer to the y-layer is bottom-up processing. Propagation in the opposite direction is top-down processing. This bidirectional processing carries out the mutual transformation between the upper level concepts and the lower level values. This is the feature ii) of the FAMOUS. The bidirectional processing interprets human macro qualitative representation into micro representation. The micro representation is available for refining the knowledge. This is the feature iii) of the FAMOUS.

3 Facial expressions recognition and generation

A facial expression model which recognizes and generates facial expressions has been proposed using FAMOUS [5][6]. The model employs the two features previous noted: inductive learning and bidirectional processing. Inductive learning enables model constructed by instances, if the fuzzy rule cannot be explicitly described. Bidirectional processing enables the model to recognize and generate facial expressions by means of the same architecture. The upper half of Figure 3 displays the facial expression model. The nodes in the concept layer indicate facial expression concepts. The nodes of the second layer from the top indicate generalized symbols obtained from instances of inductive learning for each facial feature: eyebrows, eyes, and mouth. Facial expression is separated into three aspects because it consists of pattern combinations of these different parts. The third layer from the top consists of nodes representing fuzzy labels which correspond to the abstract amount of facial characteristic movement. The node in this third layer is called the meaning element. The meaning elements have three types of fuzzy labels: small, medium, and big.

In the recognition process, inputs are given to the meaning elements. Even though the inputs are lacking and/or vague, the combination of bottom-up processing and top-down processing recalls patterns appropriate to different contexts. Here, the contexts are the memorized patterns in the model. The activation values of nodes in the concept layer display the recognition results. In the generation process, one node in the concept layer is activated, reverberation occurs and the activations are propagated. Finally, the meaning elements are activated. The meaning element has a membership function and the activation value is transformed into a numerical value through defuzzification. Numerical value refers to a concrete amount of facial characteristic movement. The values are used in order to generate an image of a facial expression.

The facial expression model can be refined through linguistic instruction learning [6]. This form of learning can be understood to occur on the conceptual level. The bottom of Figure 3 shows the instruction interpreter. Here, linguistic instructions are

5

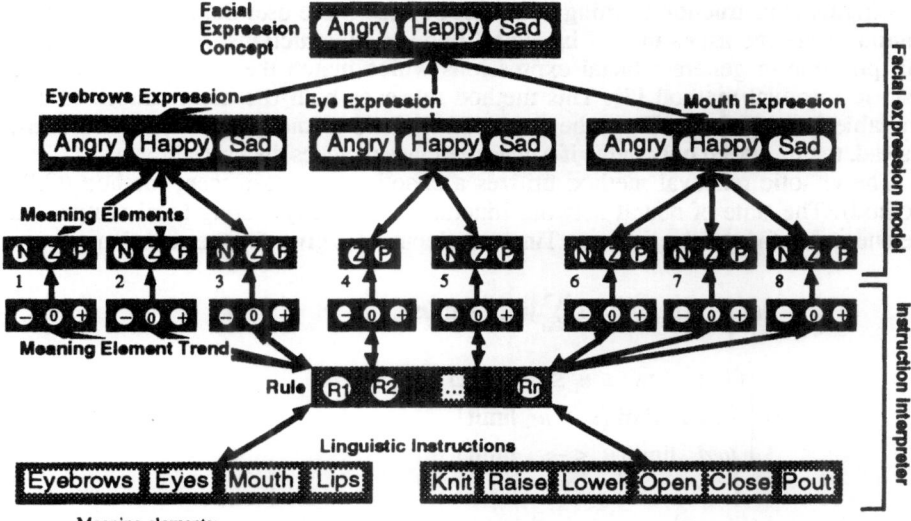

Meaning elements:
1: Distance between eyebrows, 2: Movement of eyebrows center, 3: Movement of eyebrow edges,
4: Movement of upper eyelids, 5: Movement of lower eyelids, 6: Width of mouth,
7: Movement of upper lip, 8: Movement of lower lip.

Fig. 3 A facial expression model and an interpreter of linguistic instruction

transformed into amounts of modification in meaning elements. The interpreter
consists of three layers: instructions, rules, and meaning element trends.
The instruction layers are given linguistic instructions, such as "raise eyebrows". The
rule layer transforms these instructions into meaning element trends. Each trend layer
corresponds to a meaning element layer in the facial expression model. The direction
of change in a meaning element is represented by the distribution of node activation
in the trend layer. There are three types of change: increase (+), decrease (-), or no
change (0). For example, when the distance between eyebrows is increased by the
acceptance of an instruction, the (+) node has the biggest activation value within the
trend of eyebrow distance layer. Image (a) in Figure 4 is the facial expression image
before learning. Image (b) is after learning. The user can refine the model by
linguistic instruction to obtain a model most closely approximately the user's mental
image.

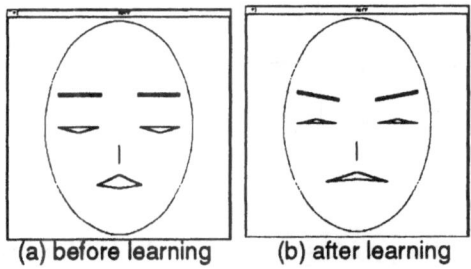

Fig. 4 Angry faces generated by the facial expression model.

6

Linguistic instruction learning could not be used if the user did not know how to communicate the user's mental image by linguistic instructions. Nevertheless, it is still possible to generate facial expressions which match the user's image by the chaotic retrieval method [7]. This method retrieves both the non-memorized and available facial patterns. Thus, the user does not have to know linguistic instructions. Instead, the user need only judge if a generated face matches the user's image.

The chaotic retrieval method utilizes a chaotic steepest-descent method (CSD method). The state of neural network itinerates chaotically among local minima on the energy surface of the network. Time development is given by the equation,

$$m\ddot{u}_i + f(\dot{u}_i, \omega t) = \varepsilon \sum_j w_{ij} a_j + d_3 init\_a_i$$

$$+ \begin{cases} 0(-u\_\text{limit} \le u_i \le u\_\text{limit}) \\ -ued\_\text{limit}(u_i > u\_\text{limit}) \\ ued\_\text{limit}(u_i < -u\_\text{limit}) \end{cases} \tag{3}$$

$$f(\dot{u}_i, \omega t) = [d_0 \sin(\omega t) + d_1]\dot{u}_i + d_2 \dot{u}_i^2 sgn(\dot{u}_i) \tag{4}$$

The second term on the left-hand side of Eq. (3) is a nonlinear damper whose gain periodically oscillates. The u_i and a_i terms are an internal state and an output for i-th neuron. Terms m and e are positive coefficients. The weight value from the j-th neuron to the i-th one is w_{ij}. The function $f(.)$ describes the nonlinear damper. w represents the angular velocity by which the nonlinear damper change gain. Term d_3 is the coefficient for an external input and $init\_a_i$ is an external input. The absolute value region of the neuron internal state is given by u_limit. The ued_limit is the revision value of the energy gradient. Terms d_0 and d_1 are coefficients for the linear damper. d_2 is the constant gain for the nonlinear damper. Finally, $sgn(.)$ is a sign function.

Chaotic retrieval for the facial expression model involves the following steps [7].
(a) A happy facial expression is generated without the CSD method.
(b) Chaotic retrieval is carried out around the patterns, adding the activation pattern for the happy facial expression to all of the nodes in the model as an external input. The Euclid distance between the external input pattern and the state of the model is calculated in each step.

Fig. 5 Happy faces retrieved by the CSD method.

(c) The model state itinerates chaotically from valley to valley on the energy surface. When the state reaches the bottom of a valley, the corresponding facial expression is produced.

Figure 5 provides three images of chaotic retrieval for a happy face. These results indicate that this retrieval method can perform the following functions. First, memorized patterns within a limited distance from an input pattern can be dynamically retrieved. Second, non-memorized and available patterns can also be retrieved. It is also apparent that the method can generate facial expressions which more closely resemble the user's image. Because of these functions, the method seems highly effective in supporting creative thinking in image generation.

4 Constructing knowledge of physical performance estimation

People can easily estimate human movements using macro qualitative language. For example, sports instructors estimate physical performance of tennis and golf. The instructor experimentally captures knowledge of estimation. There are two problems in constructing this type of knowledge in an artificial machine. First, the relation between numerical values and macro qualitative language is vague. Second, the estimation results depend on the appraisers. To solve these problems, this section proposes a FAMOUS-based construction method of knowledge which estimates human movements. The feature i) is used in the method. The method also uses results estimated by human appraisers. The appraisers observe instances of human movements, estimate them, and select the most suitable linguistic label from a list of labels. A FAMOUS is constructed using instances and their estimated results.

Figure 6 shows a model which estimates human physical performances. The model consists of three types of layers: an x-layer, several r-layers, and a y-layer. The x-layer has nodes which represent fuzzy labels about the trends of meaning elements. The meaning elements (MEs) are micro representations of physical movements. The trend of an ME is fuzzified difference in the ME between a standard movement and a movement to be estimated. The nodes in the x-layer are called "trend-nodes". Several r-layers have nodes which represent instances estimated by the appraisers; these nodes are called "instance-nodes". Each r-layer corresponds to an appraiser. For example, the r(a)-layer has nodes representing instances which are estimated by an appraiser(a). The knowledge of the FAMOUS is increased by increasing the number of r-layers. The y-layer has nodes which represent linguistic labels of estimation; these nodes are called "estimation-nodes". The most activated estimation-node represents the result of inference.

A correlation matrix between the x-layer and one of the r-layers is obtained as follows. The instances are compared with a standard movement in terms of MEs, and the differences in MEs are fuzzified using membership functions. A fuzzy grade is regarded as an activation of the corresponding trend-node. These activations are elements of an activation vector in the x-layer. The activation vector in the r-layer is made by having an instance-node corresponding to the instance given activation 1 and the other nodes given activation 0s. The matrix is obtained using activation vectors of the x-layer and the r-layer by means of Eq. (2).

A correlation matrix between the y-layer and one of the r-layers is obtained as follows. The activation vector in the r-layer is given in the same way described above. In the y-layer, an estimation-node corresponding to estimated result by an appraiser is

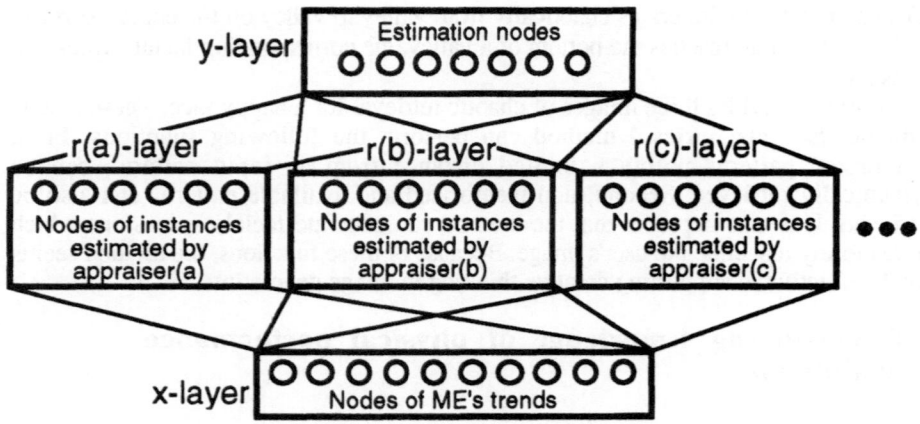

Fig. 6 FAMOUS-based knowledge representation
for estimating physical performance.

given activation 1 and the other nodes are given activation 0s. The activation vectors
of the r-layer and the y-layer are used to calculate the matrix.

The associative inference is carried out as follows. When a new instance is given
to the model, the nodes of instances which have similar features in terms of MEs
have higher activations in the r-layers. The estimation-nodes which are closely related
the activated instance-nodes have higher activations in the y-layer. Even if the
estimations by the appraisers are different in constructing knowledge, the activations
converge into the most suitable state by reverberation. The converged state reflects the
estimation results of appraisers.

5 Refining knowledge using conceptual level learning

In learning to play sports or drive a car, the students learn by means of macro
qualitative instructions which are represented by language and teaching motions.
Conceptual level learning (CLL) is proposed as a method which imitates such a
human learning process. CLL uses the FAMOUS's features ii) and iii). Being an
extension of linguistic instruction learning [8], CLL can also be used for multi-modal
instructions represented by gestures and other means, in addition to language. Figure
7 shows the image of CLL. The supervisor observes system performance, evaluates it
on several criteria, and gives instructions to it using a macro instruction represented
by words, gesture, or other means. The system interprets the macro instructions as
trends of MEs using a FAMOUS-based interpreter. It then modifies standard
knowledge represented by MEs. The FAMOUS-based CLL can be used to modify
details through macro instructions and to accommodate gestures as well as language.
This paper applies a FAMOUS-based CLL to refine MEs of standard movements
which are used in estimating human movement. The CLL discussed here consists of
linguistic and gestural instruction learning.

The CLL uses the features ii) and iii) of the FAMOUS-based knowledge. In other
words, the estimation model described in Section 4 is used as an interpreter for macro
instructions.

9

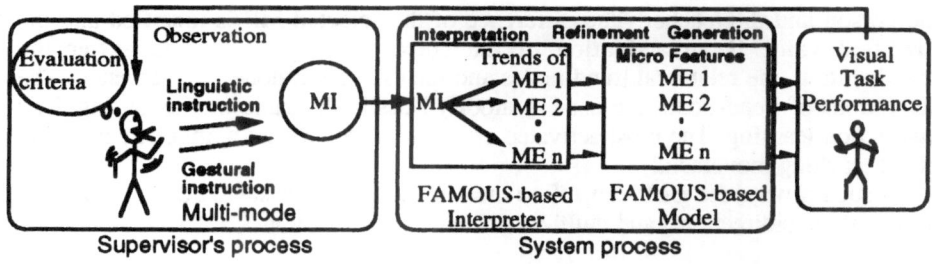

MI: Macro Instruction, ME: Meaning Element

Fig. 7 Conceptual level learning.

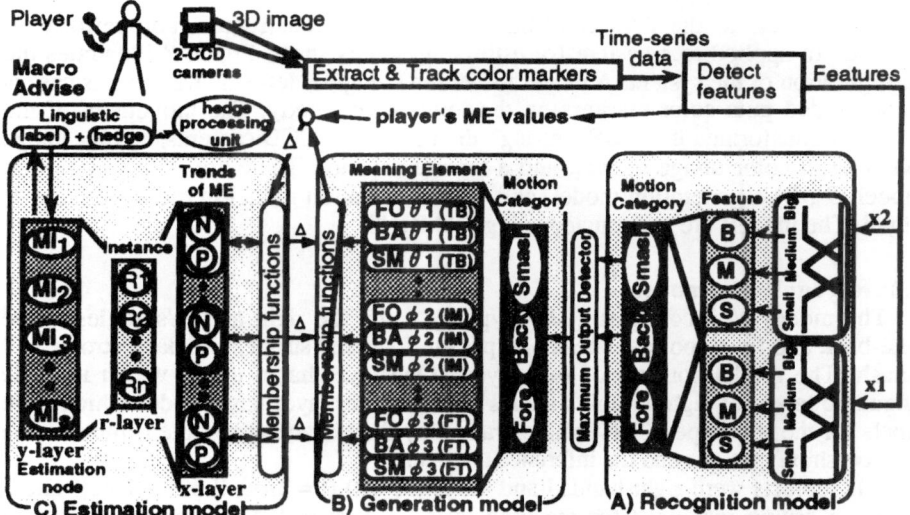

MI: Macro Instruction, ME: Meaning element, TB: Take-back, IM: Impact, FT: Follow-through,
FO: Forehand, BA: Backhand, SM: Smash,
N: Negative trend, P: Positive trend, S: Small, M: Medium, B: Big.

Fig. 8 The system for estimating human tennis shots.

In linguistic instruction learning, the estimation model transforms a supervisor's linguistic instruction into trends of MEs. Each estimation-node in the y-layer corresponds to a linguistic instruction of the supervisor. When an instruction is given to the estimation model, the corresponding estimation-node is activated. After reverberation according to Eq. (1), an activation vector of the x-layer is obtained. The activation of trend-nodes is defuzzified and the defuzzified value is used to modify the corresponding ME of a standard movement. The modification of the ME results in the refinement of the standard movement.

In gestural instruction learning, a teaching movement of a supervisor is compared with a standard movement. The differences in terms of MEs are computed and a ME which has the largest difference is selected. This process imitates the human learning process in which the student detects point which has the biggest difference between the person and the teacher. The difference of the selected ME is fuzzified and the fuzzified value is an activation of the corresponding trend-node. After the reverberation, the relational trend-nodes and the estimation-nodes are activated. The activations of trend-nodes are used to modify MEs in the same way as in linguistic instruction learning. The most activated estimation-node shows a linguistic meaning of the teaching movement.

In such a way, the CLL can refine the knowledge about standard movement by means of the bidirectional and multi-modal features of FAMOUS.

6 Tennis shot estimation system

6.1 System Configuration

In this section, the FAMOUS-based knowledge construction and refinement method are applied to a system for estimating tennis shots. Figure 8 illustrates the configuration of the system. A player located in the upper left side who in this case is right-handed puts color markers on the joints of his body. The movement of the player is transformed into MEs through the use of two CCD cameras, a color image extractor, a color image tracker, and a feature detector. There are A) a recognition model [9], B) a generation model, and C) an estimation model in the lower side of Fig. 8. The models are constructed using FAMOUS.

6.2 Recognition model [9]

The model A) recognizes a shot type by using the right armpit's angles at the take-back and finish points. The shot types are forehand stroke, backhand stroke, and smash. The model consists of two layers. One layer has 6 nodes which indicate linguistic labels of right armpit's angles and the other layer has 3 nodes which have labels for the shot types. The value extracted by the extraction module is input to the membership functions and the inference is driven.

All subjects were right handed and the angle $q(t)$ ($t = 1/60, 2/60, ...$ [sec]) under the right armpit was used for the time-series pattern in the experiment. Thus, an inflection point in the wave $q(t)$ appears when the direction of the swing of the arm changes. The top or the bottom of the point in the wave corresponds to "take-back" or "finish". The height or the depth of the point is, thus, used as characteristic value. The characteristic value of "take-back" is defined as x_1; that of "finish" is defined as x_2. There are three membership functions for each characteristic value. These membership functions are "FO" (forehand stroke), "BA" (backhand stroke), and "SM" (smash) in each point. The fuzzy rules, listed below, are embedded in the model A) as shown in Figure 8.

> Rule 1: IF x_1 is FO and x_2 is FO, THEN the motion is a forehand stroke.
> Rule 2: IF x_1 is BA and x_2 is BA, THEN the motion is a backhand stroke.
> Rule 3: IF x_1 is SM and x_2 is SM, THEN the motion is a smash.

In the experiment, six subjects swung their arms six times for each tennis motion, producing 108 samples. To examine recognition robustness with unknown individuals, the primary subject, whose data was used to obtain the membership functions, was not included in the test group. Recognition performance of the proposed system was then compared with that of conventional fuzzy inference and a

Table 1 Correct recognition ratio (%)

	FAMOUS	MLP	Fuzzy
Forehand	86.1	88.9	75.0
Backhand	77.8	77.8	72.2
Smash	88.9	69.4	66.7
Total	84.2	78.7	71.3

Number of samples : 108
FAMOUS: Inference by FAMOUS
MLP: Multi-layer perceptron trained by back-propagation
Fuzzy: Conventional fuzzy inference

three-layered perceptron. The membership functions and fuzzy rules in the conventional inference were the same as in the associative inference. The three-layered perceptron learned the ranges of the membership functions with a back-propagation algorithm. Table 1 lists ratios of correct recognition. The performance of the fuzzy associative inference system is better than that of the conventional fuzzy inference. This performance differential can be explained. When the input condition does not match a condition in a rule, conventional inference cannot select a rule or produce a result using a rule. Associative inference can overcome this handicap if another condition in the same rule is available. Activation of a node, indicates a condition is propagated from the feature layer to the motion category layer by bottom-up processing. Activation of the motion category layer is propagated to the other condition node in the feature layer by top-down processing. Associative inference recalls the most similar pattern among the memories by means of bi-directional processing. The performance of the perceptron is lower than that of the associative inference because it learned only the ranges of the membership functions. This is despite the fact that the perceptron needed considerable amounts of learning data.

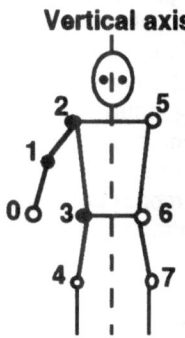

Fig. 9 Model of a human skeleton.

6.3 Generation model

The model B) generates ME's values for standard movements which correspond to the same types of shots recognized by the model A). The model B) consists of two layers. The motion category (MC) layer has three nodes representing tennis shots: forehand-stroke (FO), backhand-stroke (BA), and smash (SM).

The meaning element layer (ME) has several fuzzy labels defining, at special states, the polar coordinates of the black circled joints illustrated in Figure 9. Two types of polar coordinates are used for each joint. One is the horizontal direction angle ϕ and the other is the vertical direction angle θ. The length between the two joints is normalized. The special states are take-back (TB), impact (IM), and follow-through (FT). ·

The relationship between the MC layer and the ME layer is represented by these fuzzy rules:

IF θ_1(TB) is FO ... and ϕ_2(IM) is FO ... and ϕ_3(FT) is FO,
THEN motion is forehand.
IF θ_1(TB) is BA ... and ϕ_2(IM) is BA ... and ϕ_3(FT) is BA,
THEN motion is backhand.
IF θ_1(TB) is SM ... and ϕ_2(IM) is SM ... and ϕ_3(FT) is SM,
THEN motion is smash.

Each fuzzy label has a singleton membership function indicating a polar coordinate obtained from actual measurements. This membership function is called a standard membership function (SMF), because SMFs are used to generate standard shots.

The condition parts of the rules are assigned to the ME layer. The conclusion parts are assigned to the MC layer. The connection between the two layers is obtained by calculating correlation matrices.

In the ME layer, each fuzzy label, A {Pi (t) }, corresponds to a shot A (A = FO, BA, or SM), polar coordinate P (P = ϕ or θ), joint i (i = 1(right elbow), 2(right shoulder), or 3(right waist)), and special state t (t = TB, IM, or FT). The total number of MEs is 18 because there are three joints, two coordinates, and three states.

To generate motion, a node corresponding to the shot to be generated is activated in the MC layer. Reverberation, the propagation of activation values throughout the model, occurs. The distribution of the values then converges.

Polar coordinates in the form of the weighted-average in the SMFs are obtained using the activation values of the ME layer in the converged model. The standard time-series data for the corresponding motion are modified using these coordinates. As the human body is assumed to be symmetrical, the values for joint 5 can be calculated from those of joint 2. The values for joint 3 can be calculated from those of joint 6. Actual measured values are used to represent the values of joints 0, 4, and 7. Motion can thus be produced by computer animation using these polar coordinates.

6.4 Estimation model

In order to estimate a player's shot, the shot is compared with the standard shot generated by B) in terms of MEs. The differences in MEs are fuzzified using membership functions and the outputs from the functions are given to the x-layers in the model C) as activations.

The model C) has three types of layers as described in Section 4. The trend-nodes in the x-layer accept trends of MEs and the estimation-nodes in the y-layer output the linguistic labels in the cases of estimation. The estimation-nodes accept linguistic instructions and the trend-nodes output trends of MEs in the case of linguistic instruction learning. The trend-nodes accept and output trends of MEs and the estimation-nodes output linguistic labels in the case of gestural instruction learning. The corresponding instance-nodes in the r-layers are activated in all cases.

Each ME has two types of trends: a negative trend and a positive trend. If a difference in an ME is negative the trend becomes negative, and if a difference in a ME is positive the trend becomes positive. The total number of trend-nodes in the model C) is 36 because there are 18 MEs.

Linguistic labels

Linguistic labels (Table 2) are used in estimations or the CLL. Each estimation label corresponds to an instruction label one-to-one. These labels are assigned to the estimation-nodes in the model C).

Table 2 Linguistic labels.

Estimation Labels	Instruction Labels
Take-back is (small / big).	Make take-back (smaller / bigger).
Take-back is (low / high).	(Lower / Raise) the arm when take-back.
Body-turn is (small / big) when take-back.	Turn the body (smaller / bigger) when take-back.
Swing is (quick / slow).	Swing (quickly / slowly).
Hit-point is (low / high).	Hit at a (lower / higher) point.
Body-turn is (quick / slow).	Turn the body (quickly / slowly).
Follow-through is (quick / slow).	Swing (quickly / slowly) when follow-through.
Follow-through is (low / high),	Swing (lower / higher) when follow-through.
Body-turn is (small / big) when follow-through.	Turn the body (smaller / bigger) when follow-through.

Linguistic Hedges

A linguistic label cannot represent the amount of the difference in the ME. To solve the problem, linguistic hedges are used. The hedges are defined in hedge processing unit of Figure 8.

The estimation process uses three types of hedges: slightly, rather, and very. One of the hedges is selected as follows. When selected estimation-node is activated and the reverberation is driven again, the most closely related trend-node has the highest activation in the x-layer. The coordinate value of the ME which corresponds to the most activated trend-node is computed and the value is compared with that of the standard shot. The computed result is then defuzzified using membership functions of linguistic hedges. The linguistic hedge which has the highest fuzzy grade is selected as the linguistic hedge of the estimation label.

The linguistic instruction may also include linguistic hedges, such as "more". The system treats six types of linguistic hedges: non, 0.1; slightly, 0.2; rather, 0.4; more, 0.6; pretty, 0.8; very, 1.0. The linguistic hedges are followed by coefficients for modification. If the system is given a linguistic instruction, the corresponding nodes will be activated and the activation values will be propagated in it. Thus, the relational trend-nodes will have large activation values. The system modifies the SMFs according to Eqs. (5) and (6),

$$d\_mf = hedge \times (act\_v(N) \times neg + act\_v(p) \times pos)$$
$$/ (act\_v(N) \times act\_v(P)) \qquad (5)$$

$$new\_SMF = d\_mf + SMF . \qquad (6)$$

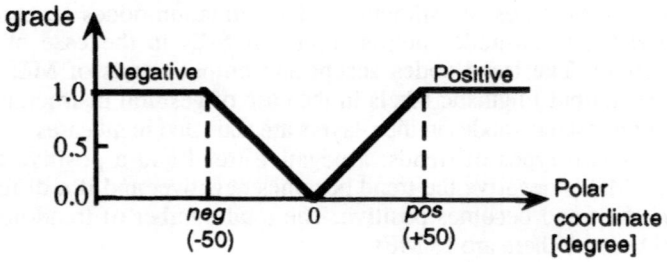

Fig. 10 Membership functions for negative and positive trends

In Eq. (5), d_mf is the amount of modification for a SMF. The term *hedge* is a coefficient for modification. $act\_v(.)$ is an activation of negative(N) or positive(P) trend-nodes. The *neg* and *pos* are values given in Figure 10. In Eq. (6), new_SMF and SMF are SMF values after and before modification.

6.5 Knowledge generalization

The method described in Section 4 requires an increased number of memories by increasing the number of appraisers. To solve this problem, we collect several r-layers into one r-layer and call this process knowledge generalization. The generalization method can be used under the precondition that all appraisers estimate the same instances completely. The method in Section 4 makes instance-nodes (R_{ia}, R_{ib}, R_{ic}, ...) for an instance i which is estimated by appraisers (a, b, c, ...) and makes weighted connections (W_{jia}, W_{jib}, W_{jic}, ...) between estimation-node E_j and the instance-nodes. The generalization method averages the weights according to Eq. (7).

$$W_{ji} = \frac{W_{jia} + W_{jib} + W_{jic} + \cdots}{N} \tag{7}$$

In Eq. (7), N is the number of appraisers. The generalization method collects nodes (R_{ia}, R_{ib}, R_{ic}, ...) into a node R_i. W_{ji} is the weights of the connection between R_i and E_j. A non-generalization method requires the same number of r-layers as the number of appraisers but the generalization method uses only one layer even if the number of appraisers increases, thus preventing the need for an increase in the number of memories.

7 Experiments

We used 40 instances of forehand stroke and 20 appraisers in the experiment. The instances and the appraisers were divided into two groups: a constructing group and a testing group. Ten appraisers estimated 20 instances each to construct knowledge of the model C) and the other 10 appraisers estimated the other 20 instances to test the constructed knowledge.

7.1 Estimation

The estimation results obtained with the FAMOUS were compared with the these of the appraisers, and the ratio of agreement between the FAMOUS and the appraisers was calculated as follows.

The FAMOUS outputted the results in a higher order of activations in the y-layer. An average of the maximum activation and the minimum activation was calculated as a threshold. The estimation-nodes with an activation higher than the threshold were selected and the estimation labels corresponding to the selected nodes were regarded as the estimation result of the FAMOUS. The agreement ratio is given by Eq. (8)

$$Ratio = Q \ / \ IN \qquad\qquad (8)$$

In Eq. (8), Ratio is the agreement ratio, Q is the number of times in which there was agreement between an appraiser and the FAMOUS, I is the total number of instances, and N is the number of appraisers. Figure 11 illustrates the ratio. In the figure, the horizontal axis is the number of appraisers used in constructing the knowledge and the vertical axis is the ratio. In tests, the ratio of non-generalization was 0.873 and the ratio of generalization was 0.811 after FAMOUS had learned the estimation result of 10 appraisers. As can be seen in the figure, the estimation result of FAMOUS converged as the number of appraisers increased. The knowledge of the model C) is constructed by instances instead of if-then rules, and the result shows that the proposed method provides the system with the function 1).

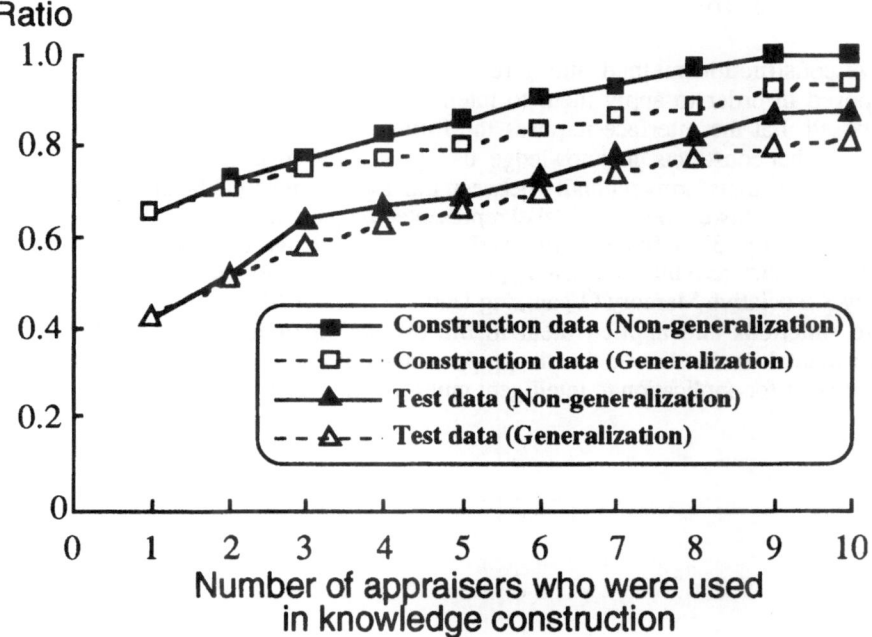

Fig. 11 Estimating tennis shots: experimental result.

7.2 Linguistic instruction learning

Figure 12 illustrates the results obtained for a shot refined by an instruction, "Lower the arm slightly when take-back". The broken line is the pre-instruction shot and the solid line is the post-instruction shot. The process refined the take-back and impact states, because this was the supervisor's macro-instruction intent.

In another experiment, the generated shots were brought close to the target shot by linguistic instruction (Fig. 13). The horizontal axis is the number of instructions and the vertical axis is the averaged errors in terms of MEs. In comparing linguistic instruction learning with micro instruction learning, a linguistic instruction modifies several related MEs at one time but a micro instruction modifies only one. As a result, the error of linguistic instruction learning decreased faster than that of micro instruction learning.

These results show that the bidirectional relations between linguistic instructions and MEs are constructed by means of instances.

7.3 Gestural instruction learning

Figure 14 shows the result of gestural instruction learning. In the figure, the broken line is the target shot and the solid line is the generated shots. In this case, the process detected the biggest difference point in the impact state and refined the related MEs according to the result of the reverberation in the model C). The process also transformed the target shot into the linguistic label "Hit at a lower point".

Functions 2) and 3) are attained in the linguistic instruction learning and the gestural instruction learning by means of the features ii) and iii).

8 Conclusion

A construction method and a refinement method of fuzzy knowledge were proposed in order to apply them to intelligent multi-modal interfaces. This paper supposed that the interface requires the following three functions at least. 1) A function that constructs the knowledge using instances instead of if-then rules. 2) A function that transforms mutually between the upper conceptual label represented by words and the lower conceptual label represented by physical values in order to realize multi-modality. 3) A function that refines the knowledge using macro qualitative instruction such as a human learning process. This paper proposed the methods using Fuzzy Associative Memory Organizing Units System (FAMOUS) in order to attain these functions and applied them to the estimation of human movements. The experimental results showed that the proposed methods provide the three functions and are suitable for application to intelligent multi-modal interfaces.

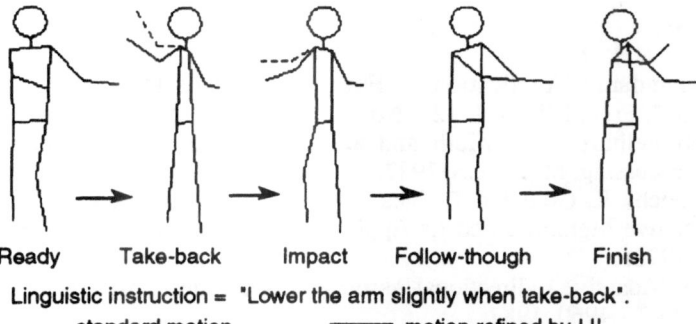

Ready　　　Take-back　　　Impact　　Follow-though　　　Finish

Linguistic instruction = "Lower the arm slightly when take-back".

－ － － standard motion　　　——— motion refined by LIL

Fig. 12 Linguistic instruction learning: experimental result.

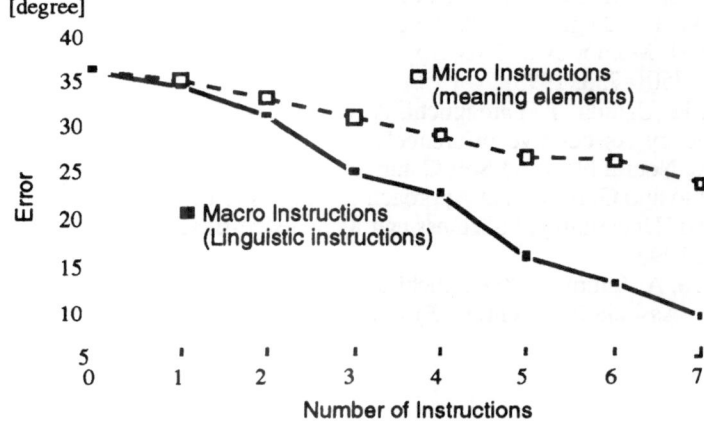

Fig. 13 Experimental comparison between macro instruction and micro instruction.

Fig. 14 Gestural instruction learning: experimental result.

References

[1] E. H. Mamdani, "Application of Fuzzy Algorithms for Control of Simple Dynamic Plant", Proc. IEE, Vol. 121, No. 12, pp. 1585 - 1588, 1974.

[2] D. E. Rumelhart, J. L. Mclleland and the PDP Research Group, Parallel Distributed Processing, MIT Press, 1987.

[3] T. Yamaguchi, K. Goto and T. Takagi, "Two-Degree-of-Freedom Fuzzy Model Using Associative Memories and Its Applications", Information Sciences, Vol. 71, pp. 65 - 97, 1993.

[4] B. Kosko, "Adaptive Bidirectional Associative Memories", Applied Opt., Vol. 26, No. 23, pp. 4947 - 4960, 1987.

[5] H. Ushida, T. Takagi, and T. Yamaguchi, "Facial Expression Model Construction Based on Associative Memories and Linguistic Instruction Learning", The Transactions of T. IEE Japan, Vol. 113-C, No. 12, pp. 1062-1071, 1993.

[6] H. Ushida, T. Takagi, and T. Yamaguchi, "Linguistic Instructions Learning Based on Associative Memories and Its Application to a Facial Model", Proc. of IJCNN Nagoya, pp. 750 - 753, 1993.

[7] T. Sato, H. Ushida, T. Yamaguchi, A. Imura, and T. Takagi, "Chaotic Memory Search in Fuzzy Associative Inference", Proc. of 3rd International Conference on Fuzzy Logic, Neural Nets and Soft Computing (Iizuka '94), pp. 203 - 206, 1994.

[8] M. Sugeno and G. Park, "An Approach to Linguistic Instruction Based Learning", Int'l Journal of Uncertainty, Fuzziness and Knowledge-Based Systems, Vol. 1, No. 1, pp. 19 - 56, 1993.

[9] H. Ushida, A. Imura, T. Yamaguchi and T. Takagi, "Human-Motion Recognition using Fuzzy Associative Memory System", Proc. of WCNN'94, Vol. 1, pp. 799 - 804, 1994.

Hybrid Connectionist Fuzzy Systems for Speech Recognition and The Use of Connectionist Production Systems

Nikola K Kasabov

Department of Information Science, University of Otago,
P.O.Box 56, Dunedin, New Zealand

Abstract: The paper presents a model for solving speech recognition tasks by exploring hybrid systems which include neural networks and fuzzy-rule based systems. The model utilises a set of neural networks for pattern recognition and a connectionist production system (CPS) for a higher level processing. Fuzzy rules for language processing are realised in the CPS. The whole process of speech recognition and language processing is considered to be one integrated process having two tightly coupled and interacting phases without a rigid, crisp border between them. The fuzziness and the ambiguity at the border line between the pattern matching and language understanding can be well represented in CPS. It facilitates flexible reasoning over fuzzy linguistic rules. An experiment on phonemes and digits recognition in English language is given for illustration. CPS make possible connectionist implementation of the whole process of spoken language recognition both at low level and at a higher level. This brings all the benefits of the connectionist systems to the practical applications in the speech recognition area.

1 Introduction

A spoken language system, as defined in [2], combines speech recognition, natural language processing and human interface technology. Spoken language systems technology has made a rapid advance in the last few years. There are now systems which work reasonably well on continuous and spontaneous speech. Some of them are speaker independent (different speakers can be recognised by the system) and adaptable (the system can adapt to new speakers if necessary). However they are limited in number and in their applications. Such systems work usually in a very restricted domain on a small (about 100 words) or medium-size vocabulary (about 1000-2000 words) with an overall correct understanding rate of about 90% [2,15,19,20,21]. The ultimate goal of building robust and adaptable speech recognition systems for continuous speech, speaker-independent, for large or non-limited vocabulary, has not been achieved so far. It is difficult to predict when it will happen, but it is clear that new techniques need to be explored [2,15]. There are several key areas of future research which have been pointed out in [2] as significant for the future development of the human computer interfaces, namely: 1) robust speech recognition; 2) automatic training and adaptation; 3) spontaneous speech; 4) dialogue models; 5) natural language response generation; 6) speech synthesis and speech generation; 7) multilingual systems; 8) interactive multimodal systems.

Hybrid systems which combine known techniques have been already tested on speech recognition tasks [4,5,15]. At a lower level, basic speech elements are recognised. Connectionist techniques have been widely used for the lower level recognition [15]. Feedforward networks with the backpropagation algorithm [4,5], Kohonen self-organising feature maps [4,15], time-delay networks [22,24-26], recurrent networks [15] and other connectionist models have been successfully applied. For the higher level recognition different techniques have been experimented, such as rule-based systems [4], Hidden Markov Models [15, fuzzy systems [5,8,9,10]. Though hybrid systems have proved to perform better than a single technique used, their achievement towards the ultimate goal has been very modest. One of the reasons is that the system should be able to deal with the ambiguity in the language at different levels of the recognition process. The system should have facilities to represent and process ambiguous, contradictory, incomplete and uncertain knowledge along the whole process of recognition.

Two are the major issues for the higher level rule-based processing, i.e.:
• *articulating the rules,*
• *interpreting the rules.*

Articulating rules for language processing, which are fuzzy by nature, is a difficult task for the language and the knowledge engineering experts. Such rules, used by humans, reflect their knowledge on the language which is not a simple 'flat' set of rules, but possibly a hierarchical structure of many *layers of rules*. These layers interact when operating before the decision on the most likely spoken phoneme, word, phrase or concept is made. They also interact with the low level processing modules. The more we learn and practice a language, the more "layers" we build in our brain and the more complex speech constructions we are able to recognise and comprehend. This is because ambiguities in the meaning of speech can occur from the acoustic right through to the syntactic and conceptual level. A higher level processing (higher than the level at which the ambiguity occurs) is required by the 'hearer' of the utterance. This is especially true for continuous, spontaneous and unlimited vocabulary speech recognition. Fuzzy systems have proved to be a suitable representation frame for speech and langauge knowledge in this respect [23]. Fuzzy rules are used in the model discussed in this paper. Having in mind the difficulties in articulating fuzzy rules for speech and language recognition we cannot overestimate the need of methods and tools for automated rule extraction and rule modification from speech and language data. Therefore a knowledge acquisition module is required as a part of a spoken language system.

Interpreting the rules is another important part of the higher-level knowledge-based system. Different methods for approximate reasoning produce different results even with the same set of rules. The approximate reasoning methods used for speech and language processing should be flexible and tunable to be able to reflect variations of accents and styles and accommodate new knowledge. The existing fuzzy inference methods have their advantages in solving control and decision making tasks and also for implementing similarity-based query systems [23,27]. But they are not flexible enough to allow for parametric tuning of the fuzzy rules without changing the membership functions of the fuzzy predicates. More flexible approximate reasoning methods are required for the purpose of higher level speech and language processing. The paper suggests using connectionist production system (CPS) for the purpose of interpreting fuzzy rules at the higher-level of speech recognition process.

The paper has the following structure. Section two presents a hybrid connectionist fuzzy model for speech recognition where neural networks are used for a low-level phoneme recognition and a fuzzy system is used for a higher-level word and phrase recognition. Section three presents a CPS called NPS (Neural Production System) from the point of view of its ability to interpret in a flexible and parametric tunable way a set of complex fuzzy rules. Section four gives and example of using NPS for a simple task of phonemes and digits recognition.

2. A hybrid connectionist fuzzy rule-based model for speech recognition

The model described here is based on the assumption that speech recognition and language modelling is an integrated continuous process and the borders between the two are not well defined, i.e. they are fuzzy. Where does the speech recognition process end and where does the language modelling process begin

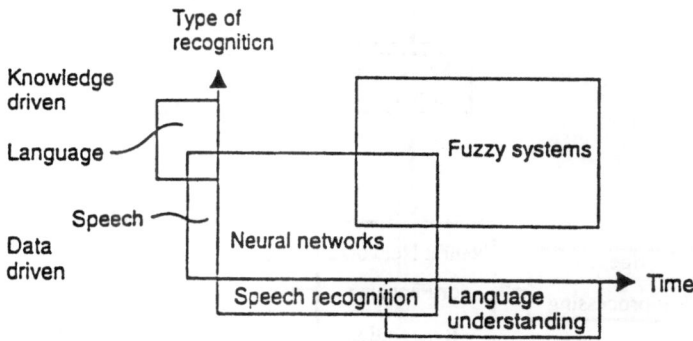

Figure 1. The two phases of speech recognition and language modelling which overlap on the "time" scale and on the "type of recognition" scale can be realised by using neural networks (low-level phase) and fuzzy systems (higher-level phase)

indeed? The approach taken here is based on using neural networks for pattern matching at a low-level. Part of the whole speech recognition process is realised at a higher, language modelling level by using fuzzy rules and approximate reasoning techniques. Two overlapping phases of recognition on the *time scale* and on *required knowledge scale* (from data driven to knowledge driven recognition) are represented by using neural networks (for the lower-level phase) and fuzzy systems (for the higher-level phase) as shown schematically in fig.1.

A hybrid system for speech recognition which uses neural networks for a low-level pattern matching and phoneme labelling and fuzzy rules for a higher level modelling is shown in fig.2. The system uses speech data corpus for training the neural networks and linguistic knowledge-base for the language modelling. A dictionary of words is also used for the language modelling. The system has a knowledge acquisition module in it for the purpose of extracting fuzzy rules form speech data in addition to priori rules which can be directly implemented in the fuzzy system.

The *low-level neural network module* can be realised as a single neural network [15], or as a multi-network structure [9]. The network recognises sub-units, e.g. portions of phones labelled by their corresponding phoneme labels [ ]. The network produces probabilities (possibilities) for a phoneme to represent a spoken signal within a small time-frame. These possibilities represent uncertainties at the current phase of recognition. These uncertainties should be processed further on when linguistic knowledge is applied. Applying linguistic knowledge is the only way to deal with ambiguities during the recognition process.

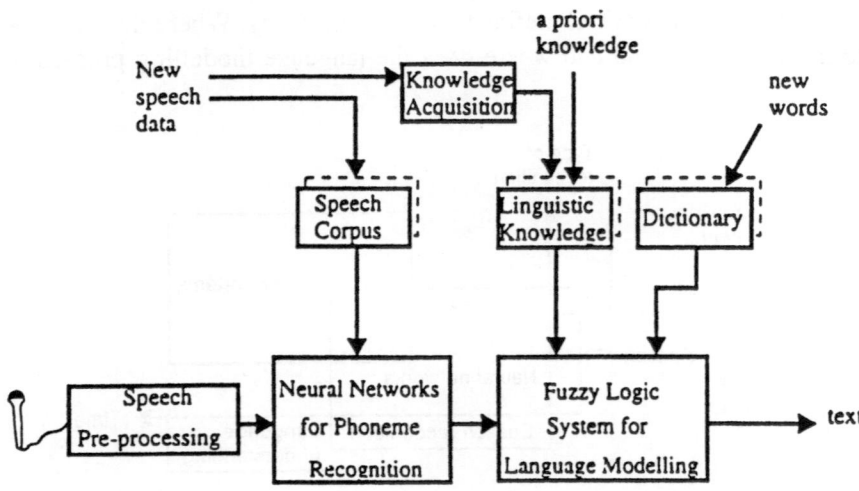

Figure 2. A hybrid system for speech recognition based on neural networks for a low-level pattern matching and fuzzy rules for language modelling

The higher-level fuzzy system module can encompass knowledge on phonetics, syntax, semantics, pragmatics. For example, ambiguity at the phoneme recognition phase can be reduced by applying linguistic rules which impose certain restrictions on the place of appearance of certain phonemes in the words [3]. Fuzzy rules can represent the likelihood of a phoneme to happen or not to happen when certain phonemes have proceeded or are following it [   ].

Different types of fuzzy rules can be articulated and used in this model, such as:

• *general acoustic fuzzy rules* which express that a phoneme is likely to happen

if at least two consecutive time-frames from the speech signal are assigned by the neural network at the lower-level recognition high values for that phoneme's label. Such a rule can have two antecedent elements A1 and A2 representing the output value produced by the network for the same phoneme label at two consecutive time moments (t-1) and (t), and a consequent - the likelihood that this phoneme has been presented in the speech signal. For example:

IF /n/$^{(t-1)}$ is Large and /n/$^{(t)}$ is Large
THEN the likelihood for phoneme /n/ is Large

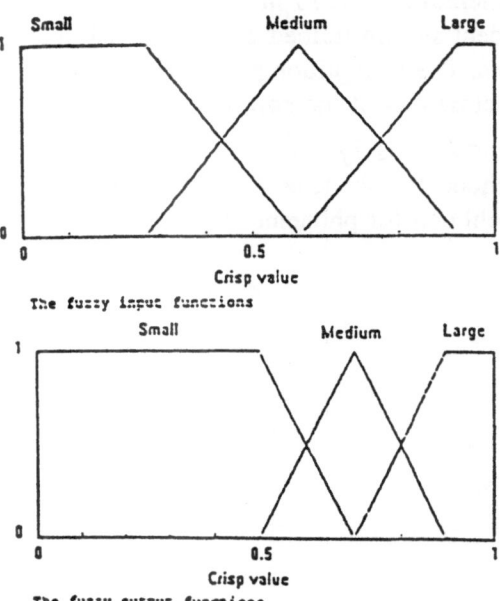

Figure 3. Exemplar membership functions for defining fuzzy predicates for input variables (these are the outputs from the previous-level neural networks) and output variable - the likelihood of a phoneme to have happened.

Similar rules can be articulated when the level of the output from the neural network is Small or Medium. One set of membership functions for the above used fuzzy terms is graphically represented in fig. 3. It would be more economical if we could represent a general rule in the form of:

IF $/?x/^{(t-1)}$ is ?y and $/?x/^{(t)}$ is ?y
THEN the likelihood of phoneme /?x/ is ?y,

where ?x is a *variable* which can be bound to any phoneme label and ?y is a variable which can be bound to any fuzzy membership label (we assume same labels used for the input and for the output fuzzy predicates even if they have different membership functions). This unfortunately is not facilitated in the fuzzy logic methods and tools. But it is possible in NPS as it will be illustrated in the next two sections.

• *specific acoustic fuzzy rules*, i.e. rules which are specific to phoneme realizations in certain accents, for example the realization of four vowels in RP, New Zealand English and general Australian English as shown in fig.4. Averaged frequencies for the first two formats of realized phonemes by a certain number of speakers are represented in two dimensional space [13].The figure shows that the phonemes / / and /I/ in New Zealand English are very close and a speaker-independent system trained on NZ English might not recognize well these two phonemes. The recognition performance may improve if there is rule which reflects the closeness of the two phonemes, e.g.:

IF / $/^{(t-1)}$ is ?y and / $/^{(t)}$ is ?y
THEN [the likelihood for phoneme / / is ?y (0.95)
    AND the likelihood for phoneme /I/ is ?y (0.7)]

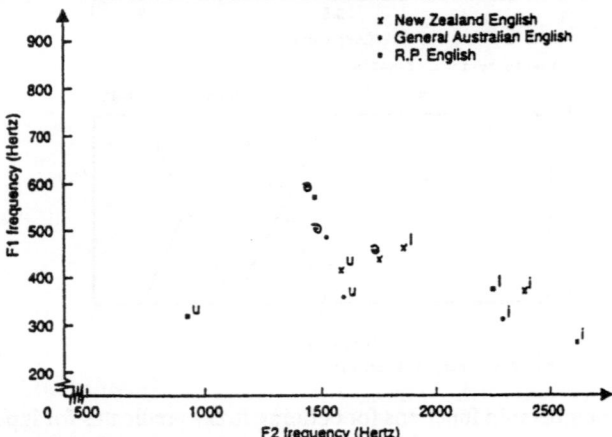

Figure 4. Mapping the two first formants representation of averaged over a number of speakers vowels in RP English, NZ English and the General Australian English. The map shows possible sources of ambiguities when realizations of different phonemes are very close to each other (redrawn from [13]).

or:
   IF / $f^{(t-1)}$ is Large and / $f^{(t)}$ is Large
   THEN [the likelihood for phoneme / / is Large
      AND the likelihood for phoneme /I/ is Medium]

Representing slight differences in pronunciation and accent requires more sophisticated types of fuzzy rules where coefficients of importance are attached to the antecedent elements and certainty factors (coefficients of belief) are attached to the consequent elements, e.g.:
   IF /n/$^{(t-1)}$ is Large (importance 1) and /n/$^{(t)}$ is Large (importance 2)
   THEN the likelihood of phoneme /n/ is Large (belief 0.9)
which gives priority to the most *recent* facts.

   More parameters, such as noise tolerance and sensitivity factors may also be needed [6,12]. Unfortunately, the standard fuzzy inference methods do not facilitate such representation. How NPS interprets more sophisticated production rules is shown in [12] and explained in the next section.

• *linguistic fuzzy rules* - these are rules which represent likelihood of certain phonemes to appear in the words of the language concerned, for example rules preventing three consonants to be recognised in a sequence, or rules which represent that after a certain phoneme has been recognized one particular phoneme is more likely to follow then another.

   The hybrid model for spoken language processing described above have the following features:

1) Mixing training data and explicit knowledge in one system - the system contains neural networks which are trained with speech and language data, and fuzzy systems, which incorporate explicit linguistic knowledge. Such systems are robust and flexible. They facilitate better use of all the sources of information available for the task [8,11].

2) Extendability - the system should allow for easy extension by adding new items to the speech corpus, adding new linguistic knowledge, adding new words to the dictionary according the concrete application.

3) Automatic training and adaptation - The system should be able to accommodate new speech data by additional training of the neural networks. It should accommodate new linguistic knowledge by adding it in a form of fuzzy rules. The system can be developed in a way to be able to automatically collect speech data from users, add them to the corpus, re-train and thus adapt to the new speakers if necessary. A challenging problem is to achieve adaptation in a real time by introducing new connectionist or hybrid connectionist-fuzzy methods. That will be an important research problem for investigation.

4) Knowledge acquisition and knowledge accumulation - The system from fig.2 facilitates 'extracting' new knowledge from new or old data through the use of the knowledge acquisition module. This will improve human's understanding on the available data from different sources of information [7,8,14,16,17].

3. Connectionist Production Systems. The NPS Realisation.

A connectionist production system (CPS) consists of a set of production rules (production memory), a set of (fuzzy) facts (working memory) and an inference machine.

The production rules (called here *generalised production rules*) implemented in a CPS NPS (Neural Production System) [6,12] are of the following form:

R_i: IF $[C_{i1} (DI_{i1}), C_{i2} (DI_{i2}),..., C_{ik} (DI_{ik})] (NT_i, SF_i)$
      THEN $[A_{i1}, A_{i2},..., A_{im}] (RF_i, CD_i)$

where: i=1,2,..n and n is the number of the rules in the fuzzy production system; C_{ij}(j=1,2,...,k) are condition elements which represent either fuzzy or boolean (crisp) propositions. They are of the form of Cij= <attribute   value>, where instead of a value, a *variable* can be used (the importance of this feature was discussed in the previous section). DI_{ij} are degrees of importance attached to the condition elements C_{ij}. A_{ij}, for j=1,2,...,m, are actions or conclusion elements of the rule. They are propositions like the condition elements in the rule which can either be added to the production memory, or retracted from it.

Two *'filtering' coefficients* represent how much existing data should match rule R_i to 'fire' this rule. The *noise tolerance* NT_i coefficient represents a level, below which any existing data is supposed to be 'a noise'. NT_i serves as a threshold above which all the data support is considered to be relevant to the rule. *Sensitivity factor* SF_i defines how sensitive the activation of a rule R_i should be depending on data support. These two coefficients define in a more sophisticated way the partial match between facts and rules.

Two coefficients, attached to every rule represent the uncertainty in the consequent part of the rule. The *reactiveness factor* RF_i defines the level of activation of the rule. The *certainty degree* CD_i defines how much the activated rule should contribute to the degree of the inferred conclusion in the right hand side of the rule.

Example: AB(7) Bx(3) Cx(1) -> Dx -Ex FC (0.5, 0.3, 1, 0.7), where: x is a variable in the rule; NT=0.5, SF=0.3, RF=1, CD=0.7 are values specified for the parameters introduced above for representing different types of uncertainty in the rule.

Each rule is represented in NPS as a node (neuron) which activation shows the *activation level* of the rule itself. *Data* in NPS is represented in the working memory as columns of neurons. Each column represents one fuzzy attribute and each neuron, its fuzzy value. Neighbouring neurons may represent lateral fuzzy values of which membership functions may overlap. The connections between the working memory neurons and production memory neurons reflect the degrees of importance in the fuzzy rules.

Each neuron *i* from the production memory, which represents a production rule R_i, receives a *support* S_i from the current data in the working memory and the variable binding space. This support is indeed *filtered* by the use of the noise tolerance NT_i and the sensitivity factor SF_i attached to this rule, *before* the net input signal I_i to the neuron i is calculated. The two coefficients NT_i and SF_i express a kind of *'meta knowledge'*, *'consciousness'*. The rationale is that different rules react differently to same data support. Fig. 5 gives a graphical representation of the influence of the noise tolerance NT_i and sensitivity factor SF_i on the net input to the neuron representing rule R_i.

The *inference machine* of NPS consists of two phases:

1) Match data against rules and calculate the activation values for the rules; partial match is achieved by using the coefficients NT_i, SF_i, RF_i and activation thresholds of the neurons in the production memory.

2) Update the working memory. Two different updating strategies are possible:

•*parallel updating*, i.e. all the rules, which have their activation values above a defined threshold Vr (as a partial case the threshold could be Vr=0) update in parallel all the neurons in the working memory.

• *a single rule updating*, i.e. one rule only updates the working memory neurons at a cycle. The winning rule neuron (the one with the highest activation value) is chosen.

The NPS architecture uses a local representation. Its block diagram is shown in Fig. 6.

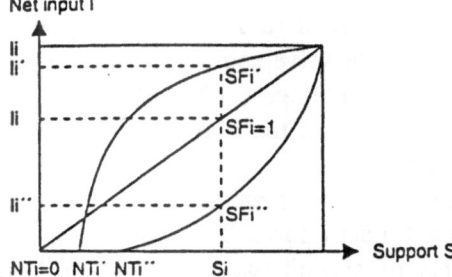

Figure 5. Different noise tolerance NTi and sensitivity factor SFi define different values for the net input Ii to a rule-neuron Ri for a same data support Si from the current facts in the working memory.

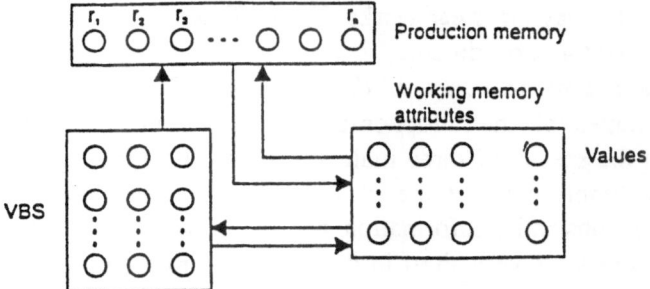

Figure 6. Block diagram of the NPS architecture

It consists of three neural sub-nets: PM - production memory; WM - working memory, and VBS - variable binding space. PM consists of n neurons, where n is the number of productions. WM is a matrix of $v\alpha$ neurons, where a is the number of the attributes and v is the number of their possible values in the condition elements of the productions. VBS is a matrix of $v.n_1$ neurons, where n_1 is the number of productions each having one variable only (at the place of the value in the pair). The variable could be a shared one between the conditions in a rule. $n_2 = n - n_1$ is the number of productions without variables. Each column of the matrix VBS corresponds to one production with a variable and each row corresponds to one of the v possible values.

Using *variables* in a production rule, which is explained in [6,12], is an interesting feature of NPS. How to make use of it for the task of speech and language recognition is still to be explored, but its usefulness is illustrated in the experiment in the next section.

4. Using NPS for approximate reasoning over speech recognition fuzzy rules

A preliminary experiment is presented here to illustrate the model introduced in section 2 and a possible use of NPS in it. The experiment is conducted on recognising eight speech units - seven phonemes and a silence as part of a system for spoken digits recognition. Eight neural networks are used to realise the low-level recognition. A set of fuzzy rules is applied on the recognised by the neural networks two consecutive segments over the time scale, labelled by the label of the corresponding phoneme. The recognized sequence of phonemes is then matched to a dictionary of the digit words. Some details on the whole experiment are given below.

• *Speech data compilation, and pre-processing [9].* The speech was recorded by using a Soundblaster card. Fast Fourier Transformations (FFT) were performed over 11.6 ms segments of speech, with the use of Hamming window and overlapping between the segments of 50%. 256 point FFTs were calculated from the speech sampled at 22,050 Hz. Mel-scaled cepstrum coefficients were then calculated from the FFTs. Mel-scale transformation is considered to be plausible in a sense with the way inner ear works. 26 coefficients are obtained from each 256 point FFT for this experiment.

• *Training the neural networks at the low-level.* To train the eight neural networks, examples of the allophonic realizations of the phonemes were extracted from the speech of three male speakers. The extracted portions were taken from the stable parts of the phonemic realisation - each of the taken segment carries substantial information about the pronounced phoneme. This meant the portions were very short in duration, but informative as explained in [9]. Mel-scale cepstrum coefficients were calculated for these portions and used to train the neural networks. Four of the neural networks for recognising four of the phonemes are shown in fig. 7.

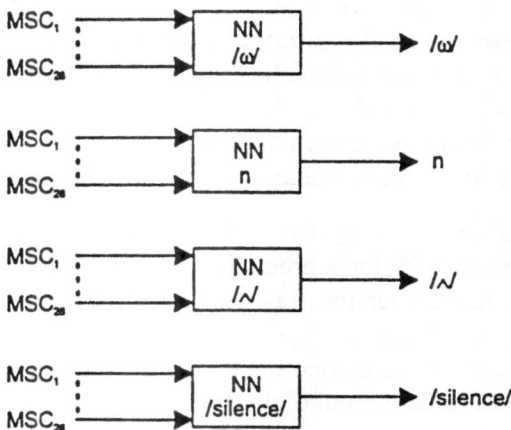

Figure 7. Four neural networks from the low-level neural network module for recognising three phonemes and the silence

• *Fuzzy rule-based system for a higher-level processing realised in NPS. General and specific acoustic rules* were articulated for deciding on the spoken phoneme at a time period based on the recognised by the neural network speech portions at shorter time moments (t-1) and (t) labelled by the label of the corresponding phoneme. The two labels are used as condition elements in rules. The first label is called "input 1" and the second - "input 2". The conclusion in a rule is called "output". Seven general and two specific rules for the case problem written in the NPS syntax are shown in fig. 8, where the following notation is used:

```
// general rules
Dx    Gx -> Ax (0.3, 0.4, 1, 0.8);
Dx    Hx -> Ax (0.3, 0.4, 1, 0.8);
Ex    Gx -> Ax (0.3, 0.4, 1, 0.8);
Ex    Hx -> Bx (0.3, 0.4, 1, 0.8);
Ex    Ix -> Bx (0.3, 0.4, 1, 0.8);
Fx    Hx -> Bx (0.3, 0.4, 1, 0.8);
Fx    Ix -> Cx (0.3 , 0.4, 1, 0.8);

// specific rules
DB DD -> AD (0.3, 0.4, 1, 0.8);
GB GD -> AD (0.3, 0.4, 1, 0.8)
```

Figure 8. General and specific rules in the NPS syntax for the task of recognition of eights spoken sub-units (for the denotation see the text). The coefficients NT, SF, RF, and CD attached to the rules are chosen through experimentation.

- for the first letter in a fuzzy proposition: A - output high; B - output medium; C - output low; D - input 1 is high; E - input 1 is medium; F - input 1 is low; G - input 2 is high; H - input 2 is medium; I - input 2 is low;
- for the second letter in the fuzzy propositions, represented as a variable x, possible values for binding it are the following phonemes: A - silence; B - /w/; C - /^/; D - /n/; E -/z/; F - /e/; G - /r/; H - /ou/.
- Same membership functions for each of the two "input" fuzzy values, different from those for the "output" fuzzy values, are used as shown in fig.3.

• *Experimenting and testing the system.* The system was tested on continuous speech. The output from NPS for a pronounced word "one" is shown in fig. 9. Before displaying, a defuzzification was done on the fuzzy values inferred by NPS. The figure shows clearly recognised sequence of phonemes which can easily be translated into the word "one". One sign denotes a level of recognition of a particular phoneme at the output of the system at a particular time frame (12 ms). Only phonemes for which the inferred degrees of recognition are greater than 0.6 are shown on the picture.

Experimental results produced by this system were compared to results obtained for the same task, same training data and same input data from a system which uses a fuzzy inference tool (TIL Shell) for realising fuzzy rules. They show an advantage of CPS when compared to the traditional fuzzy inference tools.

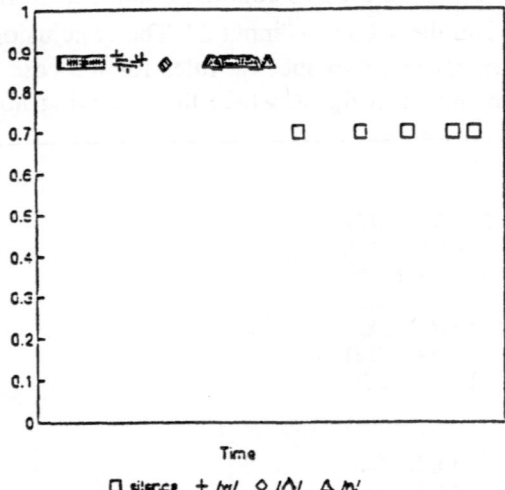

Figure 9. Recognised phonemic segments over time for a pronounced word "one". The vertical axis shows the degree of recognition of each of the phonemes at one time frame (12 msec) after defuzzification.

5. Conclusions and Directions for Further Research

A hybrid connectionist fuzzy model for building speech recognition systems is presented in the paper. It consists of pre-processing module, neural network module for low-level processing, and a higher level processing module based on fuzzy rules.

Different techniques can be used for the higher level fuzzy rules processing. Among them are the connectionist production systems (CPS). A variant of the hybrid model realisation which employs NPS as a CPS for the higher -level fuzzy rules realisation is introduced in the paper.

CPS are suitable for solving different speech and language recognition tasks. They facilitate implementing generalised fuzzy rules with coefficients of uncertainty such as degrees of importance, sensitivity factors, noise tolerance, degrees of belief. *It helps to reflect spoken language ambiguity.* They also provide a flexible and a powerful reasoning mechanism which suits the nature of the speech recognition problems. They facilitate using *generalised fuzzy variables*.

There are difficulties though in using CPS: *too many parameters to be tuned; restricted by the realisation number of variables.*

Future practical realisations of speech recognition systems are planned with the use of CPS and the proposed integrated hybrid model. Further development of CPS is also anticipated in the following directions:
- adaptability to data;
- learning rules throughout CPS' operation;
- more sophisticated updating strategies which include forgetting;
- multi-layer CPS architectures.

Acknowledgements

I would like to acknowledge the participation of Richard Kilgour, Catherine Watson and Stephen Sinclair from the University of Otago in the experiment presented in the last section. The work was partially supported by a grant from TELECOM NZ, Ltd. I would like to thank Associate Professor Takeshi Furuhashi from the Nagoya University, Japan for his support and help to finally see this paper published.

References
[1] W. Ainsworth. *Speech Recognition By Machine*, Peter Peregrinus Ltd, London, 1988.
[2] R. Cole et al. The Challenge of Spoken Language Systems: Research Directions for the Nineties, *IEEE Transactions on Speech and Audio Processing*, vol.3, No.1, January 1995, pp.1-21, 1995.

[3] A.C. Gimson. *An Introduction to the Pronunciation of English*, Edward Arnould, 1989.

[4] N. Kasabov, D. Nikovski, E.Peev. Speech recognition based on Kohonen Self Organizing Feature Maps and Hybrid Connectionist Systems, in: *N.Kasabov (ed) Artificial Neural Networks and Expert Systems*, IEEE Computer Society Press, Los Alamitos, pp. 113-117, 1993.

[5] N. Kasabov and E.Peev. Phoneme Recognition with Hierarchical Self Organised Neural Networks and Fuzzy Systems, in: *M.Marinaro and P.Morasso (eds) Proceedings of the ICANN'94*, Springer Verlag, vol.1, pp.201-204, 1993.

[6] N. Kasabov, S.Shishkov. A Connectionist Production System with a Partial Match and Its Use for Approximate Reasoning. *Connection Science*, vol.5, 3&4, pp.275-305, 1993.

[7] N. Kasabov. Learning Fuzzy Production Rules for Approximate Reasoning with Connectionist Production Systems, in: *S.Gielen and B. Kappen (Eds) Proceedings of the International Conference on Artificial Neural Networks ICANN'93*, Amsterdam, September, pp.337-345, Springer Verlag, 1993.

[8] N. Kasabov. *Neural Networks, Fuzzy Systems and Knowledge Engineering*, to be published, MIT Press, Cambridge, Massachussets, 550 pages.

[9] N. Kasabov, C.Watson, S.Sinclair, R.Kilgour. Integrating neural networks and fuzzy systems for speech recognition, in: *Proceedings of the Speech Science and Technology Conference SST-94*, University of Western Australia, Perth, Australia, December, pp. 462-467, 1994.

[10] N. Kasabov. Towards Using Hybrid Connectionist Fuzzy Production Systems for Speech Recognition, in: *Proceedings of the 1994 IEEE/Nagoya University World Wise men/women Workshop on Fuzzy Logic and Neural Networks/Genetic Algorithms*, Nagoya, Japan, August, pp. 9-13, 1994.

[11] N. Kasabov. Building Comprehensive AI and the Task of Speech Recognition, in : *J.Alspector, R.Goodman, T.Brown (eds) Applications of Neural Networks to Telecommunications*, 2, Laurence Erlbaum, 1995, pp. 178-186.

[12] N. Kasabov. Connectionist Fuzzy Production Systems. *Lecture Notes In Computer Science/Artificial Intelligence*, No.847, pp.114-132, 1994.

[13] M. Maclagan. An Acoustic Study of New Zealand Vowels, *New Zealand Speech Therapists' Journal*, vol.37, No.1, May, pp.20-26, 1982.

[14] S. Mitra and S. Pal. Fuzzy Multi-Layer Perceptron, Inferencing and Rule Generation, *IEEE Transactions on Neural Networks*, vol.6. No.1, January, pp.51-63, 1995.

[15] D. Morgan and C.Scofield. *Neural Networks and Speech Processing*, Kluwer Academic Publishers, 1991.

[16] M. Mukaidono, M. Yamaoka. A Learning Method of Fuzzy Inference Rules with Neural Networks and its Application, in: *Proceedings of the 2nd International Conference on Fuzzy Logic & Neural Networks*, Iizuka, Japan, 1992, pp.185- 187, 1992.

[17] K. Nakamura, T.Fujimaki, R. Horikawa, Y.Ageishi: Fuzzy Network Production System. In: *Proceedings of the 2nd International Conference on Fuzzy Logic & Neural Networks*, Iizuka, Japan, 1992, pp.127-130.

[18] F.J. Owens. *Signal Processing of Speech*, MacMillan, 1994.

[19] D. Pallet et al. Benchmark tests for the DARPA spoken language program, in: DARPA Workshop Speech, Natural Language Processing, March, 1993.

[20] L. R. Rabiner. Applications of Voice Processing to Telecommunications, *Proceedings of the IEEE*, vol 82, No. 2, Feb, p199-228, 1994.

[21] D.B. Roe and J.G.Wilpon. Whither Speech Recognition: The Next 25 Years, IEEE Communications Magazine, November, pp.54-62, 1993.

[22] H. Sawai, A. Waibel, P. Haffner, M. Miyatake, and K. Shikano. Parallelism, Hierarchy, Scaling in Time-Delay Neural Networks for Spotting Japanese Phonemes/CV-Syllables, in: *Proceedings of IJCANN '89*, vol. 2, pp. 81-88, 1989.

[23] T. Terano, K.Asai, M.Sugeno. *Fuzzy Systems - Theory and Applications*, Academic Press, 1992.

[24] K. Unnikrishnan, Hopfield, J.J. and Tank, D.W. Connected-Digit Speaker-Dependant Speech Recognition with Time-Delayed Connections. *IEEE Transactions on Signal Processing*, 39, 698-713, 1989.

[25] A. Waibel, T. Hanazawa, Hinton, G, Shikano, K. and K.J.Lang. Phoneme Recognition Using Time-Delay Neural Networks. *IEEE Transactions on Acoustics, Speech and Signal Processing*, 37, pp. 328-339, 1989.

[26] A. Waibel. Consonant Recognition by Modular Construction of Large Phonemic Time-Delay Neural Networks. In: *D.Touretzky (ed.) Advances in Neural Information Processing 1*. San Mateo, CA: Morgan Kaufmann, 1989.

[27] Workshop on Fuzzy Database Systems and Information Retrieval, Proceedings of the FUZZ-IEEE/IFES'95 Conference, Yokohama, Japan, March 1995.

Fuzzy Gaussian Potential Neural Networks Using a Functional Reasoning

Mohammad Teshnehlab* and Keigo Watanabe**

*Graduate School of Science and Engineering
Saga University, 1 Honjo-machi, Saga 840, Japan
**Department of Mechanical Engineering
Saga University, 1 Honjo-machi, Saga 840, Japan

Abstract – This paper presents the principal design of a fuzzy gaussian potential neural network (FGPNN) to achieve high capability to learn expert control rules of the fuzzy controller. In this construction, each membership function consists of a gaussian potential function (GPF) which causes the utilization of a reduced number of labels, and eventually the complexity of structural design becomes simple, specially for large scale inputs. This in turn reduces the learning trials, to improve the learning speed. Thus, the time of the training process, which is based on the back-propagation method, is shortened. The construction of an FGPNN is carried out with the minimum number of GPF, based on the number of input patterns, to learn the mean vectors and shapes of the individual GPFs that basically depend on the desired trajectory. Finally, we provide a simulation to evaluate the proposed method for a multi input-output, two-link manipulator.

1. Introduction

The main idea of fuzzy sets is due to Zadeh [1]-[3]. The key idea is to develop a framework to deal with imprecision. Instead of using the ordinary concept of set inclusion, Zadeh introduced a function that expressed the degree of belonging to a given set as a function taking values in the range 0 to 1. In fuzzy logic the operation corresponds to the minimum or maximum values of the membership function. Recently, fuzzy control using the fuzzy reasoning has been estabilished as one of the control technologies. As an example, any control designer can realize the fuzzy controller with minimum information of the control object, compared with the conventional model-based controls, and it can be applied to any plant, irrespective of linearity or nonlinearity of the plant.

One of the problems with fuzzy controller is the need of more trials and errors compared to other designs of optimal controllers. Another problem is to scale the input data for the fuzzy support set and to scale the inferred consequent. Another problem is to pick up the tuning of the form of membership functions and to select the control rules precisely. The determination of the number of rules from the use of the number of fuzzy labels assigned to each input is a sophisticated procedure, and in this way an effective fuzzy controller is difficult to design, because the number of control rules exponentially increases with the increase in the number of input data.

To overcome the control rule problem, self-organizing fuzzy controllers have been proposed by different authors. In this approach, to evaluate the control performance, control rules are modified by using the fuzzy reasoning. Specially, a new idea which is a combination of neural network (NN) theory and the fuzzy reasoning sounds more attractive. Such fuzzy neural networks (FNNs) [4]-[8] are applied to design controllers. In some sense the FNN automatically tunes the fuzzy rules and membership functions by using the learning function of neurons. Most of the structures of NN developed to date have focused on the update of learning parameters for a fixed NN structure.

However, if an NN has flexible function to achieve high capability in learning algorithm, then such an NN may becomes optimal network, in the sense of minimizing the control or modelling performance error.

This article presents a fuzzy gaussian potential neural network (FGPNN) to minimize the number of labels in the antecedent part of a fuzzy reasoning, which turns out to decrease the number of control rules. Thus, the proposed method becomes more effective for multi input-output systems to reduce the complexity of the architecture and hence the learning process will be improved.

This paper is organized as follows. Section 2 explains the functional reasoning. Section 3 describes the construction of membership functions in gaussian types. Section 4 explains the construction of fuzzy neural network and also describes the design of conclusion part. Section 5 describes the learning mechnisms of FGPNN and introduces the adaptation method of input scalers. Section 6 gives the dynamic equations of two-link manipulator. Section 7 gives the simulation example to evaluate the proposed FGPNN and finally the conclusion follows in Section 8.

2. Funtional Reasoning Theory

The conventional functional reasoning or its modified version has been applied to many problems. For n input variables $(x_1, .., x_n)$ and p output variables $(u_1, .., u_p)$ as the consequent, any i-th control rule can be given by

$$
\begin{aligned}
&R_i : \text{If } x_1 = A_{i1} \text{ and } \cdots \text{ and } x_n = A_{in} \\
&\quad \text{then } u_1 = f_{i1}(x_1, ..., x_n) \text{ and } \cdots \\
&\quad \text{and } u_p = f_{ip}(x_1, ..., x_n)
\end{aligned} \tag{1}
$$

where R_i is i-th control rule; A_{ij} is the fuzzy set in the antecedent associated with the j-th input variable at the i-th fuzzy rule, $f_{ik}(x_1, .., x_n)$ is the function associated with the k-th variable in the conclusion at the i-th control rule. By applying n confidences, $\mu_{Ai1}(x_1), .., \mu_{Ain}(x_n)$, the confidence in the antecedent h_i is defined by

$$
h_i = \mu_{Ai1}(x_1).\mu_{Ai2}(x_2), ..., .\mu_{Ain}(x_n) \tag{2}
$$

where "." denotes the algebraic product. Therefore, the k-th output consequent can either be derived as the weighted mean of $f_{ik}(.)$ with respect to weight h_i:

$$
u_k^* = \frac{\sum_{i=1}^{r} h_i f_{ik}}{\sum_{i=1}^{r} h_i}, \quad k = 1, ..., p \tag{3}
$$

or the weighted sum for rule's output

$$u_k^* = \sum_{i=1}^{r} h_i f_{ik} \qquad (4)$$

where r in both equations (3) and (4) is the number of fuzzy rules; if the number of membership functions in the antecedent is l, then in general $r = l^n$.

3. Construction of Membership Function

3.1 The Gaussian Type Membership Function

Horikawa et al. [4] proposed the FSNN (fuzzy sigmoid neural network) for a case to provide a pseudo-trapezodial membership function, in which we must combine two sigmoid unit functions with the ranges of $[0, 1]$ and of $[-1, 0]$. The difficulty is arised on construction method even for single input output system and small number of labelling. Watanabe et al. [7] proposed FGNN (fuzzy gaussian neural network) which is much simpler than FSNN of Horikawa et al. [4]. Also, Ichihashi [8] used similar gaussian membership function, and proposed some block hierarchical fuzzy models to reduce the number of parameters when the number of input becomes very large.

3.2 Construction of Gaussian Potential Membership Function

The gaussian potential function (GPF) used in this study consists of the squared norm of difference of input and mean value vectors . In this method, no matter when the number of inputs increases the number of fuzzy rules does not increse, but in the case of Watanabe et al. [7] the number of fuzzy rules grows with the increase of the number of inputs, thus, the structures of Watanabe et al. [7] and Horikawa et al. [4] become very complicate for the case of multi input-output systems. The proposed method with GPF turns out to become more flexible than the traditional studies by other researchers. Now, let the input data vector $x \triangleq [x_1, ..., x_n]^T$ be decomposed into some subdata vectors, e.g., $x_1 \triangleq [x_1, ..., x_m]$ and $x_2 \triangleq [x_{m+1}, ..., x_n]^T$. The gaussian potential function for the vector data x_1 can be witten by

$$F(x_1) = \exp\{-\frac{1}{2}\|x_1 - \bar{x}_1\|_{\Sigma_1^{-1}}^2\} \qquad (5)$$

where $\bar{x}_1$ is the mean value vector of x_1 and Σ_1 is the associated covariance matrix. Figure 1 shows the construction of the membership function using a neural network. In this figure, the variable with the curled bracket denotes a signal passing through the neural network, w_s is the scaler for the scalar input variable x_i, and the connection weights w_c and w_d denote the center value and the reciprocal value of the standard deviation for a gaussian function on the standardized support set.

In addition, the unit with symbol -1 generates the output of -1, the unit with the symbol Σ outputs the summation of the inputs, and the input-output relation at the unit with the symbol f is defined by

$$f(x) = x^2 \tag{6}$$

Furthermore, the unit with symbols Σ and s outputs the signal through the function s represented as

$$s(x) = e^{\ln(0.5) \cdot x} \tag{7}$$

by summing the input signals to the unit. Consequently, by using such a neural network, a GPF may be generated such as

$$s(x_1, x_2) = e^{\ln(0.5)\Sigma_{i=1}^2 [w_{si} x_i - w_{ci}]^2 w_{di}^2} \tag{8}$$

where a case with $w_{s1} = w_{s2} = 1, w_{c1} = \bar{x}_1, w_{c2} = \bar{x}_2, w_{d1} = \frac{1}{\sigma_1}$ and $w_{d2} = \frac{1}{\sigma_2}$ is shown in Fig. 1.

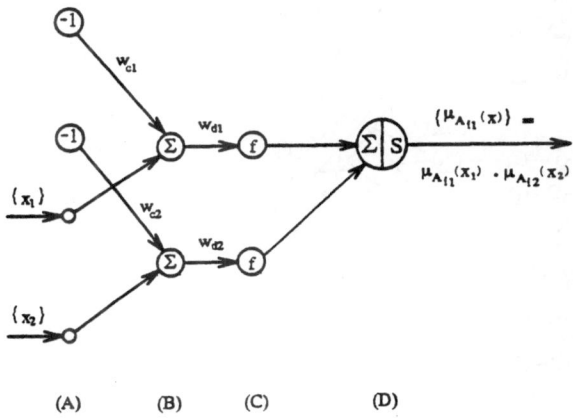

Fig. 1 Fuzzy gaussian potential membership function realized by a neural network

Using this representation gives the i-th gaussian potential membership function for the data x_1 such as

$$\mu_{Ai1}(x_1) = \exp\{-\frac{1}{2}\|x_1 - \bar{x}_1^i\|^2_{\Sigma_1^i{}^{-1}}\} \tag{9}$$

where $\bar{x}_1^i$ denotes the mean value vector of gaussian potential membership function at the i-th fuzzy rule and Σ_1^i is the associated covariance matrix. Note here that when $x_1 = \{x_1, x_2\}$ equation (9) can be rewritten by

$$\mu_{Ai1}(x_1) = \mu_{Ai1}(x_1)\mu_{Ai2}(x_2) \tag{10}$$

if scalar data x_1 and x_2 are each independent, where

$$\mu_{Ai1}(x_1) = \exp\left[-\frac{1}{2}\frac{(x_1 - \bar{x}_1^i)^2}{\sigma_{i1}^2}\right] \tag{11}$$

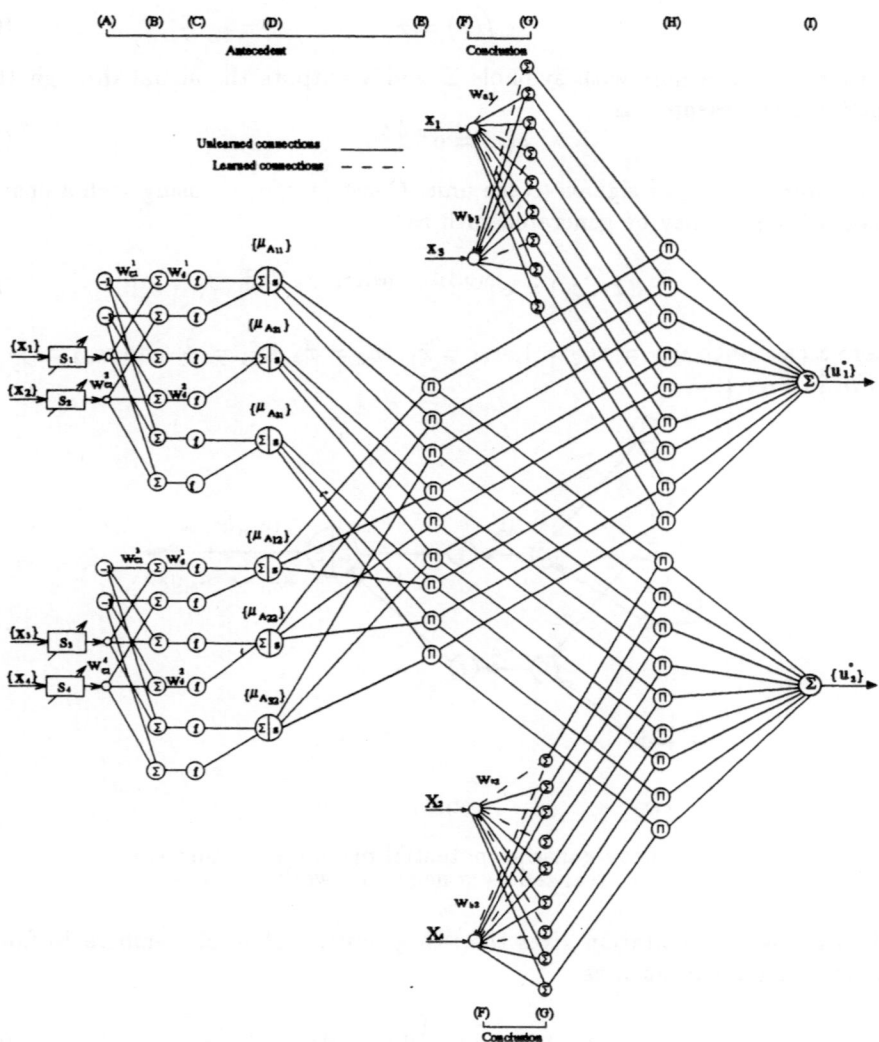

Fig. 2 Fuzzy gaussian potential neural networks
based on functional reasoning

$$\mu_{Ai2}(x_2) = \exp\left[-\frac{1}{2}\frac{(x_2 - \bar{x}_2^i)^2}{\sigma_{i2}^2}\right] \tag{12}$$

and

$$\Sigma_1^i = \mathrm{diag}(\sigma_{i1}^2, \sigma_{i2}^2). \tag{13}$$

Thus, in this approach the number of fuzzy rules does not increase even if the number of inputs becomes large, because the fuzzy labeling is made on the vector data. On the other hand, in the traditional approaches [4], [7] the number of fuzzy rules exponentially grows as the number of inputs increases, because the fuzzy labeling is made on the each scalar data. Hence, the proposed method gives a significant simplification in the FNN structure.

4. Construction of Fuzzy Gaussian Potential Neural Networks

Figure 2 illustrates the design example of an FGPNN for a case of four inputs $(x_1, ..., x_4)$ and two outputs (u_1^*, u_2^*) with three labels in the antecedent part. According to the number of labels, the number of the identifiable control rules is $r = 3^2$ if two inputs vector data are defined as $x_1 = (x_1, x_2)$ and $x_2 = (x_3, x_4)$. It should be noted that the number of identifiable control rules is $r = 3^4 = 81$ if the fuzzy labeling is made on each input data $x_1, ..., x_4$. Layers A~E correspond to the antecedent part, and layers F and G correspond to the conclusion part. The fuzzy consequent is obtained at layers H and I. At the layer C, the function f of unit is represented by (6). At the layer D, the function s of unit is written by (7). Note here that the mean values $\bar{x}_1^i, .., \bar{x}_4^i$ on the data $(x_1, ..., x_4)$ are realized as connection weights between layers A and B, and the reciprocal values $1/\sigma_{i1}, ..., 1/\sigma_{i4}$ on the deviations $(\sigma_{i1}, ..., \sigma_{i4})$ are also realized as connection weights between layers B and C.

Now, consider a case of linear function with subvector data $x_1 = (x_1, x_2)$ and $x_2 = (x_3, x_4)$ in (1). As usual applications, we suppose the tracking control problem under the condition of $x_1 = e_1$, $x_2 = e_2$, $x_3 = \dot{e}_1$ and $x_4 = \dot{e}_2$, where e_1 and e_2 denote the tracking errors of plant output 1 and 2, respectively, and $\dot{e}_1$ and $\dot{e}_2$ denote their rates of errors. If we use PD-type controllers as the conclusion functions f_{ij}, $j = 1$ and 2, it follows that

$$f_{i1} = [K_{pi}^1 \ \vdots \ 0 \ \vdots \ K_{di}^1 \ \vdots \ 0]\begin{bmatrix} e_1 \\ e_2 \\ \dot{e}_1 \\ \dot{e}_2 \end{bmatrix} \tag{14}$$

$$f_{i2} = [0 \ \vdots \ K_{pi}^2 \ \vdots \ 0 \ \vdots \ K_{di}^2]\begin{bmatrix} e_1 \\ e_2 \\ \dot{e}_1 \\ \dot{e}_2 \end{bmatrix} \tag{15}$$

for any i-th rule. Here, gains K_{pi}^1, K_{di}^1, K_{pi}^2 and K_{di}^2 can be realized as connection weight parameters in the conclusion part of an FNN to be learned.

5. Learning Mechanisms

The interesting point is to construct an FGPNN that gives a desired level of performence, even though the network structure does not have enough units. We describe the so-called back propagation algorithm, widely used algorithm for training the multilayered hierarchical neural networks. Here we consider the multilayered hierarchical neural network consisting of M layers. Let us denote the input-output relation of any unit by $f(\cdot)$, the input to the j-th unit at the k-th layer by i_j^k, and the corresponding output from its unit by o_j^k. The weight that connects the j-th unit at the k-th layer and the l-th unit at the $(k+1)$-th layer is written by $w_{j,l}^{k,k+1}$. In addition, Case A denotes a case when the input to the k-th layer is output through the function $f(\cdot)$ and the input to the $(k+1)$-th layer is calculated by the summation (i.e., Σ) operation. Similarly, Case B denotes a case when the input to the k-th layer is output through the function $f(\cdot)$ and the input to the $(k+1)$-th layer is calculated by the product (i.e., Π) operation. Note here that the output layer is assumed to consist of Case A. Under these preparations, we have the following input-output relation of a unit:

$$i_l^{k+1} = \sum_j w_{jl}^{k,k+1} o_j^k, \quad o_l^{k+1} = f(i_l^{k+1}) \tag{16}$$

for Case A and

$$i_l^{k+1} = \prod_j w_{jl}^{k,k+1} o_j^k, \quad o_l^{k+1} = f(i_l^{k+1}) \tag{17}$$

for Case B.

Now, provided a set of the input-output pattern on a plant, let us denote the teaching signal to the i-th unit of the output layer by t_i and the output from the neural network to which the input pattern is fed by o_i^M.

5.1 Generalized Learning

For the generalized learning architecture, we consider the following cost performance:

$$J = \frac{1}{2} \sum_{i=1}^{L} (t_i - o_i^M)^2 \tag{18}$$

so that the weights $w_{ij}^{k,k+1}$ minimize the above J. Here, L is the unit number of the output layer. Following a gradient descent algorithm, the increment of $w_{ij}^{k,k+1}$ denoted by $\Delta w_{ij}^{k,k+1}$ becomes

$$\Delta w_{ij}^{k,k+1} = -\eta \frac{\partial J}{\partial w_{ij}^{k,k+1}} \tag{19}$$

where $\eta > 0$ is a learning rate given by a small positive constant.

At the output layer M, since

$$\frac{\partial J}{\partial w_{ij}^{M-1,M}} = \frac{\partial J}{\partial i_j^M} \frac{\partial i_j^M}{\partial w_{ij}^{M-1,M}} = \frac{\partial J}{\partial i_j^M} o_i^{M-1} \tag{20}$$

if defining

$$\delta_j^M = -\frac{\partial J}{\partial i_j^M} \tag{21}$$

gives

$$\delta_j^M = -\frac{\partial J}{\partial o_j^M}\frac{\partial o_j^M}{\partial i_j^M} = (t_j - o_j^M)f'(i_j^M) \tag{22}$$

where $f'(i_j^M) = df(i_j^M)/di_j^M$. Therefore, the increment of the weight at the output layer is obtained by

$$\Delta w_{ij}^{M-1,M} = \eta\delta_j^M o_i^{M-1} \tag{23}$$

On the other hand, at any two intermediate layers k and $k+1$, we have

$$\frac{\partial J}{\partial w_{ij}^{k-1,k}} = \frac{\partial J}{\partial i_j^k}\frac{\partial i_j^k}{\partial w_{ij}^{k-1,k}} = \frac{\partial J}{\partial i_j^k}o_i^{k-1} \tag{24}$$

Defining

$$\delta_j^k = -\frac{\partial J}{\partial i_j^k} = -\frac{\partial J}{\partial o_j^k}\frac{\partial o_j^k}{\partial i_j^k} = -\frac{\partial J}{\partial o_j^k}f'(i_j^k) \tag{25}$$

yields

$$\frac{\partial J}{\partial o_j^k} = \sum_l \frac{\partial J}{\partial i_l^{k+1}}\frac{\partial i_l^{k+1}}{\partial o_j^k} = -\sum_l \delta_l^{k+1}w_{jl}^{k,k+1} \tag{26}$$

for Case A, and

$$\frac{\partial J}{\partial o_j^k} = \sum_l \frac{\partial J}{\partial i_l^{k+1}}\frac{\partial i_l^{k+1}}{\partial o_j^k} = -\sum_l \delta_l^{k+1}w_{jl}^{k,k+1}(\prod_{i\neq j} w_{il}^{k,k+1}o_i^k) \tag{27}$$

for Case B. Therefore, it is found that

$$\delta_j^k = f'(i_j^k)\sum_l \delta_l^{k+1}w_{jl}^{k,k+1} \text{ for case A} \tag{28}$$

$$\delta_j^k = f'(i_j^k)\sum_l \delta_l^{k+1}w_{jl}^{k,k+1}(\prod_{i\neq j} w_{il}^{k,k+1}o_i^k) \text{ for case B} \tag{29}$$

In addition, we have the increment of the weight at the intermediate layer such as:

$$\Delta w_{ij}^{k-1,k} = \eta\delta_j^k o_i^{k-1} \text{ for case A} \tag{30}$$

$$\Delta w_{ij}^{k-1,k} = \eta\delta_j^k o_i^{k-1}(\prod_{l\neq i} w_{lj}^{k-1,k}o_l^{k-1}) \text{ for case B} \tag{31}$$

Note here that, for the feedback error learning, if the j-th feedback input is represented by u_{fj}, then δ_j^M in (22) is replaced by

$$\delta_j^M = u_{fj} f'(i_j^M) \tag{32}$$

5.2 Specialized Learning

Furthermore, for specialized learning, we consider the following cost performance:

$$J = \frac{1}{2} \sum_{i=1}^{m} (y_{di} - y_i)^2 \tag{33}$$

to obtain the weights $w_{ij}^{k,k+1}$ that minimize J. Here, m denotes the number of the plant outputs, y_{di} the i-th desired reference and y_i the i-th output of the plant. Then, a further deforming of (21) gives the delta quantity at the output layer for a case of multi input-output with coupling input, described by

$$
\begin{aligned}
\delta_j^M &= -\frac{\partial J}{\partial i_j^M} \\
&= -\frac{\partial J}{\partial o_j^M} \frac{\partial o_j^M}{\partial i_j^M} \\
&= \sum_{i=1}^{m} -\frac{\partial J}{\partial y_i} \frac{\partial y_i}{\partial o_j^M} f'(i_j^M) \\
&= f'(i_j^M) \sum_{i=1}^{m} (y_{di} - y_i) \frac{\partial y_i}{\partial u_j} ,
\end{aligned}
\tag{34}
$$

where the $\partial y_i / \partial u_j$ is defined as

$$\frac{dy_i}{du_j} = \frac{\partial y_i}{\partial u_j} + \sum_{l \neq j} \frac{\partial u_i}{\partial u_l} \frac{du_l}{du_j} \tag{35}$$

$$\frac{\Delta y_i}{\Delta u_j} \simeq \frac{\partial y_i}{\partial u_j} + \sum_{l \neq j} \frac{\partial y_i}{\partial u_l} \frac{\Delta u_l}{\Delta u_j} \tag{36}$$

also u_j denotes the j-th input to the plant and the $f'(i_j^M)$ is equal one. Finally, the update equation of learning parameters are described by

$$w_{ij}^{k-k,k}(t+1) = w_{ij}^{k-1,k}(t) + \eta \delta_j^k o_i^{k-1} + \alpha \Delta w_{ij}^{k-1,k}(t) \tag{37}$$

for adjusting the shape of gaussian potential function in the antecedent and connection weights in the conclusion part, where α denotes the momentum coefficient with the range in $0 \leq \alpha < 1$

Table 1 Actual values of manipulator parameters

	Link one	Link two
Mass m_i[kg]	5.0	5.0
Inertia I_i[kg m^2]	0.104	0.104
Length l_i[m]	0.5	0.5
Length l_{gi}[m]	0.25	0.25

5.3 Adaptation of Input Scalers

Generally, the constant input scalers can be easily determined from the maximum value of the data obtained through the control experiments. But most of the times the input data fail to transform into the predetermined support set, because of the addition of the fuzzy compensator to the control system. To overcome this problem, we use a simple adaptive method for scaling the input data $x_1, ..., x_n$, as described by

$$w_{s,new} = \begin{cases} \dfrac{0.9 \times L}{|x_i|} & \text{if} \quad |w_{s,old} \times x_i| > L \\ w_{s,old} & \text{otherwise} \end{cases} \tag{38}$$

which means that if the scaled input data are scaled out from the support set $[-L, L]$, then the input data are rescaled to fall into the 90% range of the support set.

6. Manipulator Model

There has recently been a considerable interest in developing efficient control algorithms for robot manipulators. The complexity of the control problem for manipulators arises mainly from that of manipulator dynamics itself. The dynamics of articulated mechanisms in general, and of robot manipulators in particular, involve strong coupling effects between joints as well as centrifugal and Coriolis forces. The equations of motion for 2-DOF planar manipulator can be written in the compact form as

$$M(\theta)\ddot{\theta} + V(\theta, \dot{\theta}) = \tau \tag{39}$$

where $\tau \in R^2$ is a vector of joint torques supplied by the actuators, $\tau = [\tau_1 \quad \tau_2]^T$ and $\theta \in R^2$ is a vector of joint positions with $\theta = [\theta_1 \quad \theta_2]^T$, and the $M(\theta) \in R^{2\times2}$ is called manipulator inertia mass matrix, whose elements are obtained by

$$M_{11} = I_1 + I_2 + m_1 l_{g1}^2 + m_2[l_{n1}^2 + l_{g2}^2 + 2l_{n1}l_{g2}\cos(\theta_2)] \tag{40}$$

$$M_{12} = M_{21} = I_2 + m_2[l_{g2}^2 + l_{n1}l_{g2}\cos(\theta_2)] \tag{41}$$

$$M_{22} = I_2 + m_2 l_{g2}^2 \tag{42}$$

Here I_j is the moment of inertia for the j-th link, l_{gj} is the distance from the joint j to the center of gravity of the j-th link. The actual values for the parameters of the manipulator are given in Table 1.

The vector $V(\theta, \dot{\theta}) \in R^2$ represents forces arising from the Coriolis force and centrifugal force which are expressed as:

$$V_1 = -m_2 l_{n1} l_{g2} \sin(\theta_2)(2\dot{\theta}_1 + \dot{\theta}_2)\dot{\theta}_2 \tag{43}$$

$$V_2 = m_2 l_{n1} l_{g2} \sin(\theta_2)\dot{\theta}_1^2 \tag{44}$$

7. Simulation Example

We use the 4th-order Runge-Kutta-Gill method to simulate the plant dynamics state. It is assumed that the control sampling period is $T = 10$ [ms], the step width of the integration is 0.4 [ms], the initial states are $\theta_1 = 0.5236$ [rad] and $\theta_2 = 0.4363$ [rad], and $\dot{\theta}_1 = \dot{\theta}_2 = 0$. Also, the desired trajectories of the manipulator are assumed to be known as time functions of joint positions, velocities, and accelerations, that is, θ_d, $\dot{\theta}_d$, and $\ddot{\theta}_d$ are expressed as

$$\theta_{d1} = 0.5\cos(\pi t) \quad \theta_{d2} = 0.5\sin(\pi t) + 1.0 \tag{45}$$

$$\dot{\theta}_{d1} = -0.5\pi\sin(\pi t) \quad \dot{\theta}_{d2} = 0.5\pi\cos(\pi t) \tag{46}$$

$$\ddot{\theta}_{d1} = -0.5\pi^2\cos(\pi t) \quad \ddot{\theta}_{d2} = -0.5\pi^2\sin(\pi t) \tag{47}$$

The learning algorithms of both connection weights in the antecedent and conclusion parts of FGPNN are implemented. In this study five lables are applied to each input subvector as indicated in Fig. 3. The center position values, w_c, are $-4, -2, 0, 2, 4$, and the reciprocal values of standard deviation w_d are all 1, in accordance with all of the labels on the support set $[-4, 4]$. The initial values of the connection weights (i.e., PD gain parameters) in the conclusion parts are given in Table 2.

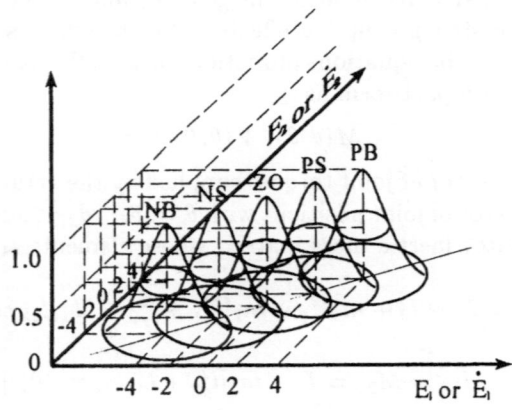

Fig. 3 The gaussian potential membership function with
5 labels on the support set [-4, 4]

45

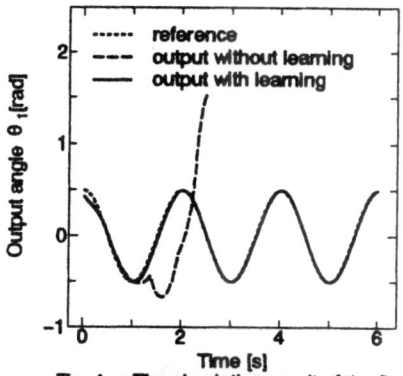

Fig. 4–a The simulation result of the first
link using proposed FGPNN controller

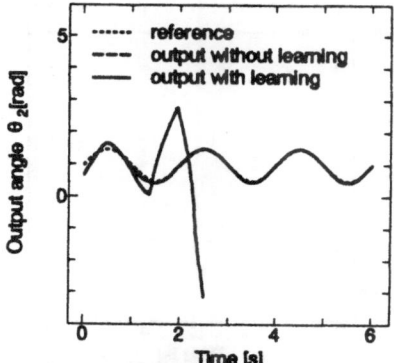

Fig. 4–b The simulation result of the second
link using proposed FGPNN controller

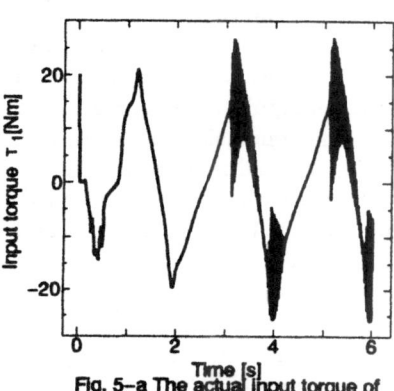

Fig. 5–a The actual input torque of
the first link

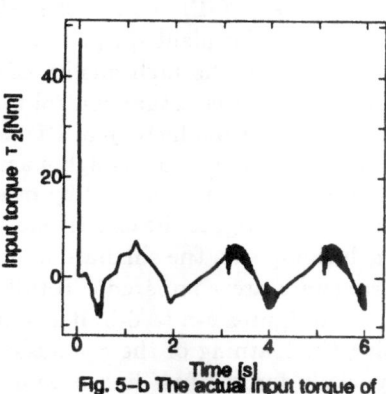

Fig. 5–b The actual input torque of
the second link

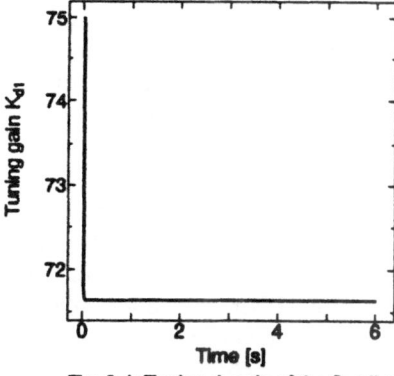

Fig. 6–b Tuning d–gain of the first link

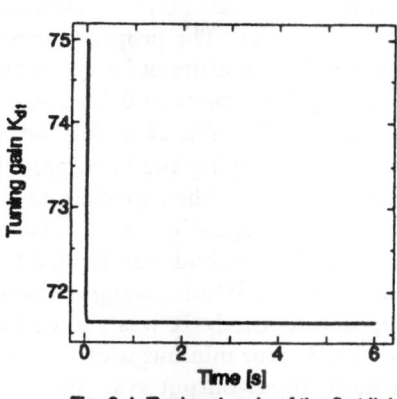

Fig. 6–b Tuning d–gain of the first link

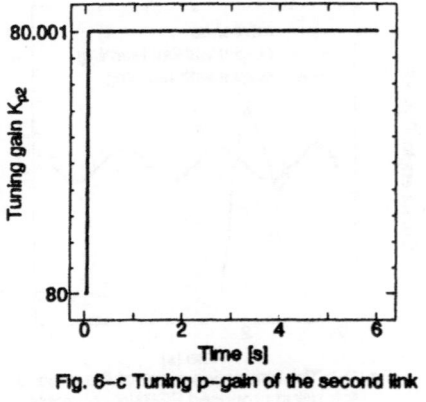

Fig. 6–c Tuning p–gain of the second link

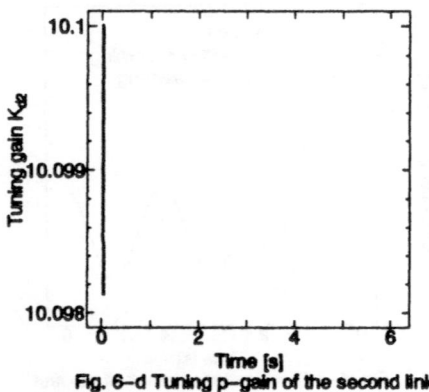

Fig. 6–d Tuning p–gain of the second link

The control results are shown in Figs 4-a and 4-b. These figures indicate that the proposed FGPNN controller shows high ability for the multi/input-output system and the plant outputs try to follow the desired trajectories which can be counted for the high quality of performance in this method. Also, Figs. 5-a and 5-b show the actual control inputs of two-link manipulator. Figures 6-a to 6-d show the time history of PD learning gains in the conclusion parts. In this study, those connection weights are indicated in Fig. 2, used in learning process and the results are replaced by other connection weights which are not learned. Also, to investigate the effectiveness of learning parameters in the antecedent and conclusion parts, the simulations with learning parameters and without learning parameters are compared. Totally, from simulation results of Tables 3-a and 3-b, and figures 6-a to 6-b, it is found that the control was mainly improved by using the learning of the conclusions, because the PD gains in conclusion parts especially for the first link gave a wide range of variations, but the parameters in the antecedent parts made slight changes.

8. Conclusion

In this paper, an FGPNN has been proposed especially for designing a learning fuzzy controller. The proposed method can be utilized for large scale systems to simplify the strcuture of fuzzy-neural network configuration by using the concept of GPFs. The input and its mean were considered to be a vector form, thus, the membership functions formed a potential form. The learning method was based on modifying the back-propagation algorithms for the purpose of on-line learning (the so-called specialized learning method). In this approach, both the connection weights in the antecedent and conclusion parts of FGPNN could be learned. The method was applied to construct a controller for robotic manipulators. The simulation example was also given to demonstrate the validity of the proposed method. It was shown that the suggested FGPNN controller gave a very good error minimization and a quick convergence to the desired trajectories for multi input-output systems.

REFERENCES

[1] L. A. Zadeh, "Fuzzy sets," *Information and Control*, Vol. 8, pp. 330-353, 1965.

[2] L. A. Zadeh, "Fuzzy algorithms," *Information and Control*, Vol. 12, pp. 94-102, 1968.

[3] L. A. Zadeh, "Common sense knowledge representation based on fuzzy logic ," *IEEE Computer*, Vol. 10, pp. 61-65, 1983.

[4] S. Horikawa, T. Furuhashi, and Y. Uchikawa, "On fuzzy modeling using fuzzy neural networks with the back-propagation algorithm," *IEEE Trans. on Neural Networks*, Vol.3, No.5, pp.801-806, 1992.

[5] J.-S. R. Jang and C.-T. Sun, "Functional Equivalence Between Radial Basis Function Networks and Fuzzy Inference Systems," *IEEE Trans. on Neural Networks*. Vol.4, No. 1, pp. 156-158, 1993.

[6] J.-S. R. Jang, "Fuzzy modeling using generalized neural networks and Kalman filter algorithm," *Proceedings of Ninth National Conference on Artificial Intelligence*, pp. 762-767, 1991.

[7] K. Watanabe, J. Tang, M. Nakamura, S. Koga, and T. Fukuda, "A Mobile Robot Control using Fuzzy-Gaussian Neural Networks," *Proceedings of IROS' 93*, Yokohama, Japan, 26-30 July 1993, Vol.2, pp.919-925.

[8] H. Ichihashi, "Learning in Hierarchical Fuzzy Models by Conjugate Gradient Method using Back-propagation Errors," *Procs. of 1st FAN Symposium*, Oct., 25 and 26, Osaka, pp.235-240, 1991.

Recurrent Fuzzy Logic Using Neural Network

Emdad Khan   Fatih Unal

Intelligent Systems Group, Embedded Systems Division, National Semiconductor, Santa Clara, CA 95052, USA

Abstract

In this paper, a novel method is presented to combine neural nets with fuzzy logic. The combined technology is based on modified NeuFuz ([1], [2],[3]) using recurrent neural networks. The recurrent information of neural net is directly mapped to a new type of fuzzy logic, called "recurrent" fuzzy logic. Recurrency preserves temporal information and yields superior performance for context dependent applications. It also reduces the convergence time. Simulations show good improvements in accuracy and speed of convergence in pattern recognition applications.

1.0 Introduction

Fuzzy logic has been proven very successful in solving problems in many areas where conventional model based (mathematical modeling of the system) approach is either very difficult or inefficient/costly to implement. Fuzzy logic based design has several advantages including simplicity & ease in design. However, fuzzy logic design is associated with some critical problems as well. As the system complexity increases, it becomes difficult to determine right set of rules and membership functions to describe the system behavior. A significant amount of time is needed to properly tune the membership functions and adjust rules before a solution is obtained. For more complex systems, it may be even impossible to come up with a working set of rules and membership functions. Besides, once the rules are determined, they remained fixed in the fuzzy logic controller i.e controller cannot learn from experience (adaptive fuzzy system, however, provides limited learning).

Use of neural nets to learn system behavior seems to be a good way to solve above mentioned problems associated with fuzzy logic based designs. Using system's input-output data, neural nets can learn systems behavior and accordingly can generate fuzzy rules ([6], [1], [7]) and membership functions. However, processing these rules and membership functions using conventional fuzzy algorithms, in general, does not produce satisfactory solution mainly because of the heuristic nature of these algorithms. The most popular fuzzy inferencing method uses the maximum of the outputs from all rules for each universe of discourse. The most popular and effective defuzzification uses center of gravity (COG) method. These methods usually yield good solutions for relatively simpler problems. For complex problems, these heuristic based algorithms may not yield satisfactory results over a wide range. Moreover, such approaches may take unusually long time to determine the correct sizes and shapes of both input and output membership functions. Determining a good set of rules that will work well with the above mentioned inferencing and defuzzification is another big problem in conventional fuzzy design.

To alleviate these problems, we have presented *([1], [2],[3])* novel methods (NeuFuz) to automatically generate fuzzy logic rules and membership functions using neural net learning and then process these using non heuristic neural net based fuzzy logic algorithms. Such approach can produce fuzzy logic rules and membership functions to meet certain pre- specified accuracy level and can significantly simplifies the design process, reduces design time and improves performance, reliability at lower cost.

In this paper, we have introduced recurrent fuzzy logic to effectively use temporal information to provide improved solution for context dependent problems. We have presented elegant methods to generate recurrent fuzzy logic rules using the basic fuzzy rule framework presented in NeuFuz. We have used the proposed techniques in different context dependent applications and obtained very encouraging results.

2.0 Mapping Recurrent Neural Net to Recurrent Fuzzy Logic

Fig.1 shows a neural network based fuzzy system [1]. For simplicity, we are using only a 3-layered neural net to represent the learning of fuzzy rules and membership functions (of a 2-input, one output system). As shown in the figure, the 1st layer neurons do fuzzification, the 2nd (or hidden) layer neurons form the rule base and the 3rd layer neuron does the rule evaluation and defuzzification.

2.1 Recurrent Neural Net

Fig. 1 shows the feedforward (non recurrent) neural nets used in NeuFuz([1], [2],[3]) to generate membership functions and non recurrent fuzzy rules. The net uses multiplying neurons in layer 2. Backpropagation algorithm is accordingly modified for multiplying neurons and used for the learning process. To handle temporal information, recurrent connections are added in layer 2 neurons as shown in Fig. 2. Thus, in calculating the output of a neuron in layer 2, outputs of layer 1 neurons connected to layer 2 neurons are multiplied with the feedback signals of layer 2 neuron with appropriate delays. Thus, we have,

$$\text{netp}_j^{\text{hidden}}(t) = (\Pi \, W_{ij} \cdot o_i) \cdot (\Pi \, \text{netp}_j^{\text{hidden}}(t - m) \cdot WR_j, m) \ldots\ldots\ldots\ldots\ldots\ldots(1)$$

where

$\text{netp}_j^{\text{hidden}}(t - m) = \text{netp}_j^{\text{hidden}}$ with delay m with respect to current time "t"

$\text{netp}_j^{\text{hidden}}(t) = $ layer 2 (hidden layer) output at time t and

$O_i = $ output of the i-th neuron in layer 1

We have used linear neurons with a slope of unity for the middle (layer 2) and output (layer 3) layer neurons. Thus, the equivalent error at the output layer is

$$d_k^{\text{out}} = (t_k - o_k) \, f'(\text{net}_k) \qquad \ldots\ldots\ldots\ldots\ldots\ldots\ldots\ldots\ldots\ldots\ldots\ldots\ldots\ldots\ldots (2)$$

where o_k is the output of the output neuron k

t_k is the desired output of the output neuron k

$f'(\text{net}_k)$ is the derivative which is unity for layers 2 & 3 neurons as mentioned above.

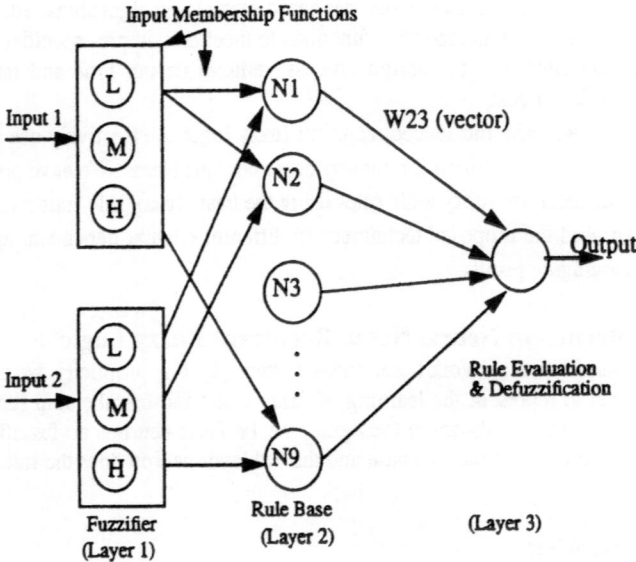

Fig. 1: A feed forward neural net to learn system behavior. The net is first trained with
system input-output data. Learning takes place by appropriately changing the
weights between the layers. After learning is completed, the final weights represents the rules and membership functions. The learned neural net, as shown above,
can generate output very close to the desired outputs. Equivalent fuzzy design
can be obtained by using generated recurrent fuzzy rules and membership functions as
described in sections 2.2.

The general equation for the equivalent error at the hidden layer neurons using Back Propagation
model is

$$d_j^{hidden} = f'(net_j) \, \Sigma \, d_k^{out} . W_{jk} \dots\dots\dots\dots\dots\dots\dots\dots\dots\dots\dots\dots(3)$$

However, for the fuzzification layer in fig.1, the equivalent error is different as for the middle
layer we have product (instead of sum) neuron. Thus, the $netp_j$ is [equation (1), rewritten here]

$$netp_j^{hidden}(t) = (\Pi \, W_{ij} . o_i) . (\Pi \, netp_j^{hidden}(t - m). WRj,m) \dots\dots\dots\dots(4)$$

Thus, for the input layer (Fig.1) the equivalent error expression (after incorporating the effect of
the recurrent paths) becomes ([1],[2])

$$d_i^{input} = f'(netp_i) \, \Sigma \, [d_j^{hidden} . W_{ij} . (\Pi \, W_{kj} . o_k) . \Pi \, netp_j^{hidden}(t - m). WRj,m] \dots.(5)$$

where both i & k are indices in the input layer and j & m are the indices in the hidden
layer.

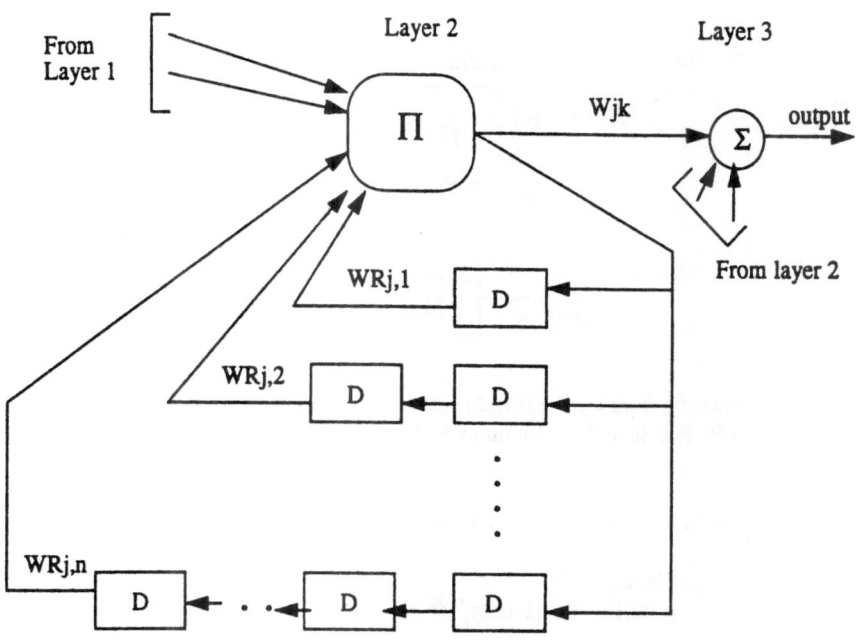

Fig. 2: Layer 2 of Fig.1 is modified with self feedback with delays. Only one neuron is shown for simplicity. D represents a unit delay. Weight WRj,n represent feedback weight of j-th neuron in layer 2 with n delays. The recurrent weights are modified during the learning process using back propagation learning as described in section 2.1. Recurrent weights are also properly used in recall mode calculations.

The weight update equation for any layer (except recurrent path in layer 2) is

$$\Delta W_{ij} = \varepsilon. \, d_j.o_i \quad\dotfill(6)$$

where ε is the learning rate.

For the feedback connections (layer 2), o_i is denoted by $o_i^{recurrent}$ and is the $netp_j^{hidden}$ (t-m) multiplied by WRj,m (which corresponds to the weight in the path with "m" delays). We add a bias term of unity to $Oi^{recurrent}$ to help convergence.

Thus, $o_i^{recurrent}$ is given by

$$o_i^{\,recurrent} = 1 + netp_j^{hidden}(t - m). \, WRj,m \quad\cdots\cdots\cdots(7)$$

And the d_j is the equivalent error in the hidden layer (equation 3).

2.1.1 Weight Update for Recurrent Path

The weight update equation for the recurrent path is somewhat different. For simplicity, consider unit delay (i.e $m = 1$). For better clarification, the neuron in layer 2 (fig. 2) is redrawn showing 2 parts: P1 and P2 (fig. 3).

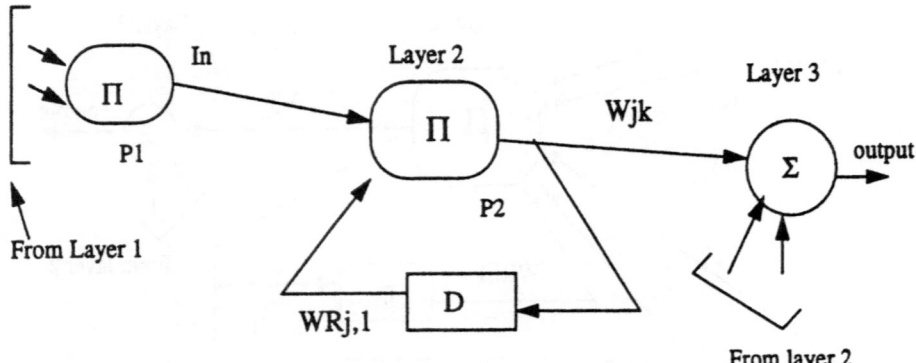

Fig. 3: A neuron from layer 2 (Fig.2) is redrawn showing 2 parts: P1 and P2. The output of P1 is is In and the output of P2 is netpj^{hidden} as used in Fig.2.

The output of P1 is In and the output of P2 is $netp_j^{hidden}$ as described before. Thus, we have,

$$netp_j^{hidden}(t) = In.\ O_i^{recurrent}$$
$$= In.(1 + WRj,1.\ netp_j^{hidden}(t-1)) \ \dots\dots\dots\dots\dots\dots (8)$$

Let $fn = netp_j^{hidden}$. Then, the partial derivative of fn with respect to WRj,1 becomes:

$$fn'(t) = In.[fn(t-1) + WRj,1.\ fn'(t-1)] \ \dots\dots\dots\dots\dots\dots(9)$$

Using standard gradient descent and chain rules, it can easily be shown that

$$\Delta WRj,1 = \varepsilon.\ d_j^{hidden}.\ fn'(t) \ \dots\dots\dots\dots\dots\dots\dots\dots(10)$$

2.1.2 Temporal Information and Weight Update

It is to be noted that the temporal information is accumulated by the recurrent path. Hence, update needs to be done in batch mode. Thus, above weight changes are calculated for each pattern (and stored) but weight update is done only after all patterns are applied and by summing all ΔWij's for each pattern. The whole cycle is then repeated until the convergence is obtained. Thus, we follow schemes similar to real-time learning as reported in [8]. Accumulation of information is also takes place in the "recall" mode, and it has corresponding impact on the fuzzy rules as described below. It should also be noted that the net only remembers the recent past i.e it gradually forgets the old context.

2.2 Recurrent Fuzzy Logic

The 1st layer neurons in Fig.1 include the fuzzification process whose task is to match the values of the input variables against the labels used in the fuzzy control rule. The 1st layer neurons and the weights between layer 1 and layer 2 are also used to define the input membership functions. In fact, it is difficult to do both fuzzification and learning membership functions just by using one layer of neurons, and we actually have used 4 layers for this ([1], [2],[3]). The layer 2 neurons represent fuzzy rules as mentioned before. The rule format for the layer 2 neurons of Fig. 1 is

"If input 1 is Low AND input 2 is Low
    THEN output is X"   [standard fuzzy rule]

where X is either a number or a linguistic variable. We have used X as a number. Thus, instead of using an output membership function, we have used singletons. These singletons are also learned by the neural net.

However, considering the recurrent connection of Fig.2, the fuzzy rule format is accordingly changed to incorporate the recurrency. Thus, the recurrent fuzzy rule format is

"If input 1 is Low AND input 2 is Low
    AND Previous output is Y1
        THEN the new output is X" [recurrent fuzzy rule with one delay]

considering recurrent connection with unit delay. Y1 is a singleton like X. However, it can have two forms:
a) Y1 same as the WRj,1 and
b) Y1 as the product of $netp_j^{hidden}(t-1)$ and WRj,1.

Clearly, the previous output information is incorporated in the rule's antecedent and is represented either by weights WRj,1 or by $(WRj,1 . netp_j^{hidden}(t-1))$, the latter though makes more sense from temporal information point of view. Thus, fuzzy processing is different for these two cases although both would essentially yield the same results.

For recurrent connection with n-delays, the recurrent fuzzy rule becomes,

"If input 1 is Low AND input 2 is Low
    AND last output is Y1
    AND 2nd last output is Y2
    .....
    .....
    AND n-th last output is Yn
        THEN the new output is X" [recurrent fuzzy rule with n delays]

To ensure that we don't lose any information (or accuracy) in mapping neural net to fuzzy logic, a one to one mapping is used (for both versions of singletons Yx's). Such approach has several advantages as explained later. To ensure one to one mapping between the neural net and the fuzzy logic, neural network based fuzzy logic algorithms are used as described below.

2.2.1 Antecedent Processing

From equations [(1), (4)] above, it is clear that our antecedent processing uses multiplication as opposed to minimum operation in conventional fuzzy logic design. Thus the equation to combine two antecedents is

$$u_c = u_a . u_b \dots\dots(11)$$

where u_c is the membership function of the combinations of the membership functions u_a & u_b.
Use of multiplication (as dictated by neural net) significantly improves the result [3].
As mentioned before, the previous output (i.e recurrent) information is used in the antecedent part of the recurrent fuzzy rule and thus, such information is processed using equation 1.

2.2.2 Rule Evaluation and Defuzzification

Consider the following equation for the proposed defuzzification:

$$\text{Output} = \Sigma \, y.Wjk \dots\dots\dots\dots\dots\dots\dots\dots\dots\dots\dots\dots\dots\dots\dots(12)$$

where Output is the final defuzzified output which include the contribution from all the rules as represented by layer 2 neurons and y represent the outputs of layer 2 neurons. Clearly, we get a defuzzification which exactly matches the neural net behavior. Hence, this defuzzification (called Neural defuzzification in [2]) is the optimal case. It is also much simpler as it does not use any division.

It is to be noted here that the defuzzification is actually rule evaluation. Since the output of the rule is a non fuzzy number, we actually don't need a defuzzification. Thus, we can call it rule evaluation rather than defuzzification.

3.0 Controlling Fuzzy Logic Accuracy

As mentioned above, using neural net based fuzzy logic algorithms, 100% mapping of neural net to fuzzy logic is obtained. Since the neural net can be trained to a pre specified accuracy level, the generated fuzzy rules and membership functions yield the same accuracy for the training set. Depending on the generalization, the unseen data may yield less accuracy. However, neural net can be trained to a higher accuracy for the training set that provides the desired accuracy for the test set. 100% mapping is guaranteed when all the generated rules are used. For smaller set of rules (after optimization) some accuracy may be lost.

4.0 Determining the Number of Time Delays

Using too many feedback delays complicate fuzzy rules, uses more memory and not necessarily improve the solution. This is because context information is useful only when the signals in the near vicinity belong to the same membership function (i.e when autocorrelation function does not decay rapidly). With large number of delays in the feedback path, all distant signals will not belong to the current membership class and so the recurrent information will not be useful. Typical values for number of delays are from 1 to 5, in our applications.

5.0 Simulations and Results

We have applied our proposed methods to different applications. The simulations showed very encouraging results. For brevity, we have presented here only a trigonometric sine wave case. and compared it with non-recurrent fuzzy design based on NeuFuz. The results are shown in Table 1. Significant improvement in learning speed and accuracy is clearly demonstrated. Clearly, context information with more than one delay does not help much.

Table 1: Number of cycles used to learn a sine wave with 9 data points.
5 membership functions were used.

Accuracy	Number of Cycles to converge				
	1 D	2 D	3 D	4 D	NeuFuz (no recurrency)
0.01	9	8	8	7	1036
0.001	172	160	160	167	5438

6.0 Summary & Conclusion

Recurrent fuzzy logic concept is introduced and an elegant method is presented to combine recurrent neural nets with recurrent fuzzy logic. Recurrent neural net is used to effectively use context information. Fuzzy logic membership functions and recurrent fuzzy rules are generated based on neural net learning. Also, the problems associated with the conventional fuzzy inferencing, defuzzification and antecedent processing methods have been addressed. Elegant methods are proposed to solve these problems of conventional fuzzy logic design by using neural net based algorithms for defuzzification, rule evaluation, and antecedent processing. Proposed recurrent fuzzy logic design significantly reduces design time with improved performance and reliability. The proposed schemes can generate fuzzy rules and membership functions to meet a pre-specified accuracy. The methods have been verified (by simulation) for several real world applications and obtained very promising results.

Future Work

Work is underway to incorporate recurrency to the inputs as well. Thus, a typical fuzzy logic rule considering single delay would look like
"If the input1 is Low AND previous input1 is Z1 AND input2 is High AND
previous input2 is Z2 and Previous output is Y1
then the new output is X".
Extension to multiple delay case is straightforward.

Re erences

[1] E. Khan NeuFuz: An Intelligent Combination of Fuzzy Logic with Neural Nets, IJCNN, Nagoya, Oct 93.
[2] E. Khan et al, "Neufuz: Neural Network Based Fuzzy Logic Design Algorithms", Proceeding of the FUZZ-IEEE93, pp 647 Vol 1, Mar 1993.
[3] E. Khan, "Neural Network Based Algorithms For Rule Evaluation & Defuzzification in Fuzzy Logic Design", to be presented at the WCNN, July 93.
[4] L. Zadeh et al, " Fuzzy Theory and Applications", Collection of Zadeh's good papers,1986
[5] D.E Rummelhart et al, "Learning Internal Representations by Error Back Propagation", Edited by James Anderson and Edward Rosenfield, MIT press, 1988.
[6], B. Kosko, " Neural nets & Fuzzy Systems", Prentice Hall, 1992
[7] Y. Hayashi et al, "Fuzzy Neural Network with Fuzzy Signals and Weights", Proceeding of IJCNN92, Vol II (Baltimore)
[8] J. Hertz et al, "Introduction to the Theory of Neural Computation", Addison-Wesley, 1991
[9] C. Lin, "Neural Network based Fuzzy Logic Control and Decision Systems", IEEE transaction on computers, Vol 40, no. 12, 1991
[10] S. Horikawa et al, "On Fuzzy Modeling using Fuzzy Neural Networks with the Back-propagation Algorithm", IEEE transactions on Neural Networks, Vol 3, no. 5 Sept 1992

Information Aggregating Networks based on Extended Sugeno's Fuzzy Integral

Keon-Myung Lee and Hyung Lee-Kwang

Dept. Computer Science, KAIST(Korea Advanced Institute of Science and Technology), Taejon, 305-701, Seoul Korea

Abstract. Sugeno's fuzzy integral is a functional to aggregate partial evaluations for an object in consideration of importance degrees of evaluation items. This paper presents the issues related to Sugeno's fuzzy integral for information aggregation. For the identification of importance degrees of evaluation items with the properties of fuzzy measures, we suggest to use a genetic algorithm based method. To improve the behavior of the fuzzy integral by avoiding excessive emphasis of pessimistic aspects, we introduce compensatory operators into the fuzzy integral. On the other hand, to tune the parameters for the used compensatory operators and to perform the fuzzy integral in parallel computation, we propose a network model.

1 Introduction

In decision making and pattern recognition, we frequently encounter the situations that we should produce an overall evaluation value from several partial evaluation values, where each partial evaluation value is determined in the viewpoint of an evaluation item. As a method to deal with these situations, there is Sugeno's fuzzy integral[6] which has the role to aggregate partial evaluation values with respect to importance degrees of evaluation items.

When we use Sugeno's fuzzy integral, we have the following problems: First, we should determine importance degrees of evaluation items holding the properties of fuzzy measure. It is not easy to provide consistent fuzzy measure values since they have to be subjectively determined. Second, Sugeno's fuzzy integral has the tendency to emphasize the pessimistic aspect by taking the minimum of evaluation values on performing fuzzy integral. Third, there is some computational burden to perform fuzzy integral since its operation is imposed on all elements of the power set of evaluation item set.

This paper proposes some solutions for the above problems. To identify fuzzy measure values for importance degree, we use genetic algorithm-based method that we have already proposed[5]. In the identification, we are interested in λ-fuzzy measure identification. To avoid excessively pessimistic evaluation, we introduce compensatory operators[11] into fuzzy integral. In addition, we extend the fuzzy integral to reflect certainty factors of evaluation values. On the other hand, to alleviate the computational burden and tune the parameters of compensatory operators, we propose an information aggregating network model based on Sugeno's fuzzy integral. To show the applicability of proposed method, we perform an experiment which evaluates preference degree for secondhand cars.

This paper is organized as follows: Section 2 describes the λ-fuzzy measure identification based on genetic algorithms. Section 3 extends Sugeno's fuzzy integral to improve its properties. Section 4 presents a network model to carry out the extended Sugeno's fuzzy integral and make it possible to tune some parameters in fuzzy integral and Section 5 shows an experiment and its results. Finally Section 6 draws conclusions.

2 Fuzzy Measure Identification

To use fuzzy integral in information aggregation, we should have importance degrees assigned to evaluation items. These importance degrees are required to preserve the properties of fuzzy measures.

2.1 Fuzzy Measures

Fuzzy measure g is a set function defined on the power set $\mathcal{B}(X)$ of X satisfying the following properties[10]:

$$g : \mathcal{B}(X) \rightarrow [0,1]$$
1) $g(\phi) = 0, \quad g(X) = 1$
2) If $A, B \in \mathcal{B}(X)$ and $A \subset B$, then $g(A) \leq g(B)$.
3) If $F_n \in \mathcal{B}(X)$ for $1 \leq n < \infty$ and a sequence
   $\{F_n\}$ is monotone (in the sense of inclusion),
   then $\lim_{n\to\infty} g(F_n) = g(\lim_{n\to\infty} F_n)$.

Among fuzzy measures, λ-fuzzy measure g_λ is a widely used fuzzy measure with the following properties:

$$\forall A, B \in \mathcal{B}(X), \ A \cap B = \phi,$$
$$g_\lambda(A \cup B) = g_\lambda(A) + g_\lambda(B) + \lambda g_\lambda(A)g_\lambda(B) \text{ for } \lambda > -1.$$

In λ-fuzzy measure for a finite set $X = \{x_1, x_2, \ldots, x_k\}$, fuzzy density values $g_i = g_\lambda(\{x_i\})$ leads the followings:

$$g_\lambda(\{x_1, \ldots, x_l\}) = \sum_{i=1}^{l} g_i + \lambda \sum_{i_1=1}^{l-1} \sum_{i_2=i_1+1}^{l} g_{i_1} g_{i_2} + \cdots$$
$$+ \lambda^{l-1} g_1 g_2 \cdots g_l = \frac{1}{\lambda}[\Pi_{i=1}^{l} (1 + \lambda g_i) - 1]$$

2.2 Genetic Algorithms

Genetic algorithms can be viewed as a general-purpose search method, an optimization method, or a learning mechanism, based loosely on Darwinian principles of biological evolution: reproduction and "survival of the fittest" along with genetic recombination[1, 2, 3, 4].

Genetic algorithms maintain a set of candidate solutions called a population. Candidate solutions are usually represented as strings of fixed length, called chromosomes. Given a (random) initial population, genetic algorithms operate in cycles, called generations, as follows:

1. Initialize a population of chromosomes.
2. Evaluate each chromosome in the population.
3. Create new chromosomes by mating current chromosomes; apply genetic operators
4. Delete some chromosomes of the population to make room for the new chromosomes.
5. Evaluate the new chromosomes and insert them into the population.
6. If time is up, stop and return the best chromosome; if not, go to 3.

When we want to use genetic algorithms for solving problems, we should develop the followings:

Encoding scheme. Each chromosome corresponds to a candidate solution and should have a format to which genetic operators can be applied. Thus we need to develop some methods to represent candidate solutions in coded strings(chromosomes). In general, chromosomes are represented by binary strings, real value strings, or symbolic strings, etc.

Initialization of population. A well-initialized population helps genetic algorithms to find desirable solution(s) more easily and faster than does a poorly-initialized one. Thus we need deliberate methods to create an initial population.

Evaluation function. During the operation of genetic algorithms, all chromosomes are evaluated to see how proper they are as solutions to the problem. Thus we have to develop a function to evaluate the fitness of candidate solutions as real solutions.

Genetic operators. Although there are typical genetic operators, we can not directly use them in our problems since they are affected by the encoding schemes and properties of problems. Thus, we must develop genetic operators suitable to the problem.

2.3 λ-Fuzzy Measure Identification with Genetic Algorithms

Fuzzy measure identification is to determine fuzzy measure values $g(A)$, $A \subset \mathcal{B}(X)$ for a set $X = \{x_1, x_2, \ldots, x_k\}$. Thus, λ-fuzzy measure identification is to decide fuzzy density values g_i, $i = 1, \ldots, k$ and λ. Since fuzzy measure values are subjectively determined, it is difficult to acquire consistent values satisfying the properties of fuzzy measures from human experts. Here we review a method to produce such fuzzy measure values from human-provided values with genetic algorithms[1, 2]. In the sequel, $\hat{g}_\lambda(A)$, $A \subset \mathcal{B}(X)$ and $\hat{g}_i$ denote the human-provided values, and $g_\lambda(A)$ and g_i denote the identified values.

The genetic algorithm for λ-fuzzy measure identification consists of two stages: Genetic Algorithm I and Genetic Algorithm II. Both stages are performed by genetic algorithms. Genetic algorithm I takes charge of determining fuzzy density values g_is, and Genetic Algorithm II finds the λ value for the given fuzzy density values g_is obtained by Genetic Algorithm I.

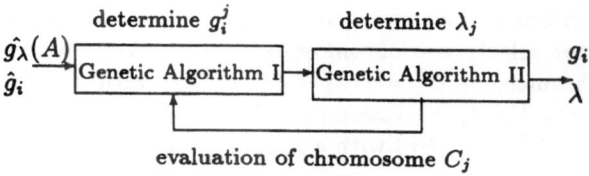

Fig. 1. λ-fuzzy measure identification genetic algorithm

In Genetic Algorithm I, chromosomes encode the fuzzy density values g_i^j and the λ value by a vector $C_j = (g_1^j, g_2^j, \ldots, g_k^j; \lambda_j)$. When $\hat{g}_i$ is available, the i-th position of each chromosome is initialized with a value near $\hat{g}_i$. When $\hat{g}_i$ is not available, the value for the i-th position is randomly selected from the interval $(0, \min\{\hat{g}_\lambda(A) \mid x_i \in A, A \in \mathcal{B}(X)\}]$.

The fitness function $f_1(C_j)$ for chromosome C_j is the sum of the differences between human-provided fuzzy measure value $\hat{g}_\lambda(A)$ and fuzzy measure value obtained by g_i^j and λ_j.

$$f_1(C_j) = \sum_{A \in \mathcal{B}(X)} |\hat{g}_\lambda(A) - \frac{1}{\lambda_j}[\Pi_{x_i \in A}(1 + \lambda_j g_i^j) - 1]|$$

In Genetic Algorithm I, there are two genetic operators : crossover and mutation. The crossover operator produces a new chromosome C' from two randomly selected chromosomes C_1 and C_2 as follows:

$$C_1 = (g_1^1, g_2^1, \ldots, g_i^1, g_{i+1}^1, \ldots, g_k^1; \lambda_1)$$
$$C_2 = (g_1^2, g_2^2, \ldots, g_i^2, g_{i+1}^2, \ldots, g_k^2; \lambda_2)$$
$$C' = (g_1^1, g_2^1, \ldots, g_i^1, g_{i+1}^2, \ldots, g_k^2; \lambda')$$

The crossover influences on only the fuzzy density values g_i^j but the λ value. In the reproduced chromosome C', λ' is determined by Genetic Algorithm II. The mutation operator selects a chromosome C_1 and a position i of the chromosome, and modifies the i-th position value with a randomly selected value r from a specified interval as follows: Here λ' is determined by Genetic Algorithm II.

$$C_1 = (g_1^1, g_2^1, \ldots, g_i^1, \ldots, g_k^1; \lambda_1)$$
$$C' = (g_1^1, g_2^1, \ldots, g_i^1 + r, \ldots, g_k^1; \lambda')$$

In Genetic Algorithm II, chromosome is a real value to represent a λ value, i.e., $C_j = (\lambda_j)$. To initialize the population, randomly select $\hat{g}_i$, $\hat{g}_j$ and $\hat{g}(\{x_i, x_j\})$, and find Λ such that $\hat{g}(\{x_i, x_j\}) = \hat{g}_i + \hat{g}_j + \Lambda \hat{g}_i \hat{g}_j$. Each chromosome is initialized with randomly selected value from the interval $[\Lambda - \Delta, \Lambda + \Delta]$ where $\Delta > 0$.

The fitness function $f_2(C_j)$ of chromosome C_j is the sum of differences between the human-provided measure value $\hat{g}_\lambda(A)$ and measure value obtained by fuzzy density values g_i produced by Genetic Algorithm I and λ_j.

$$f_2(C_j) = \sum_{A \in \mathcal{B}(X)} |\hat{g}_\lambda(A) - \frac{1}{\lambda_j}[\Pi_{x_i \in A}(1 + \lambda_j \hat{g}_i) - 1]|$$

Genetic Algorithm II uses two genetic operators: crossover and mutation. Crossover operator selects two chromosomes $C_1 = (\lambda_1)$ and $C_2 = (\lambda_2)$, and creates a new chromosome $C' = (r\lambda_1 + (1 - r)\lambda_2)$ with a randomly selected value r at the interval $[0, 1]$. Mutation operator selects a chromosome $C_j = (\lambda_j)$ and changes it into $C'_j = (\lambda_j + r)$ with a randomly selected value r at a specified interval.

3 Extension of Sugeno's Fuzzy Integral

In the literature, there are several fuzzy integrals which are a kind of Lebesque integral[10]. Among them, Sugeno's fuzzy integral is a typical one. It has the role of aggregating partial evaluations for an object in consideration of importance degrees of evaluation items.

Let X be a set of evaluation items and $g(E)$ the importance degree of evaluation item set $E \subset X$ with the properties of fuzzy measure. $g(x)$ denotes the evaluation value on the standpoint of evaluation item x, and A denotes the interest focus of evaluation items. The fuzzy integral $\oint_A h(x) \circ g(\cdot)$ over the set $A \subset X$ of the function h with respect to a fuzzy measure g is defined as follows:

$$\oint_A h(x) \circ g(\cdot) = \sup_{E \subseteq X} \{\min\{\min_{x \in E} h(x), g(A \cap E)\}\}$$

$$= \sup_{\alpha \in [0,1]} \{\min\{\alpha, g(A \cap F_\alpha)\}\}$$

$$F_\alpha = \{x \mid h(x) \geq \alpha\}$$

$$= \sup_{E \subseteq A} \{\min\{\min_{x \in E} h(x), g(E)\}\}$$

We can interpret the meaning of the fuzzy integral in the following way: In the above formula, $\min_{x \in E} h(x)$ selects the most pessimistic evaluation among the current evaluation items E. $\min\{\min_{x \in E} h(x), g(E)\}$ imposes the restriction that the aggregated evaluation value can not be greater than the importance degree of the current evaluation items E. $\sup_{E \subseteq A}\{\min\{\min_{x \in E} h(x), g(E)\}\}$ elicits the most promising evaluation value.

Due to the operation $\min_{x \in E} h(x)$, fuzzy integral has a tendency to produce pessimistic evaluation. Some decision making problems show that although an item has poor evaluation, the item can be compensated by other good items. Thus to provide the same effect to fuzzy integral, we extend it by introducing compensatory operators instead of minimum operator in the operation $\min_{x \in E} h(x)$.

There are two kinds of compensatory operators which can be used in the extended fuzzy integral: mean operators and hybrid operators.

Mean Operators

Weighted Arithmetic Mean Operator
$$A \oplus_\gamma B = (1 - \gamma)(A \cap B) + \gamma(A \cup B) \qquad 0 \leq \gamma \leq 1$$

Geometric Mean Operator
$$A \otimes_\gamma B = (A \cap B)^{(1-\gamma)}(A \cup B)^\gamma \quad 0 \le \gamma \le 1$$

Here $A \cap B$ denotes the result obtained by some t-norm operator[11] such as minimum operation, and $A \cup B$ the result by some t-conorm operator such as maximum operation. γ represents the compensation degree. Thus as γ comes to have a larger value, the compensatory operators take a more optimistic stance.

Hybrid Operators

Multiplicative γ-model
$$y = (\Pi_{i=1}^n x_i^{\delta_i})^{(1-\gamma)}(1 - \Pi_{i=1}^n (1 - x_i)^{\delta_i})^\gamma$$
$$\sum_{i=1}^n \delta_i = n, \, 0 \le \gamma \le 1$$
Additive γ-model
$$y = (1 - \gamma)\Pi_{i=1}^n x_i^{\delta_i} + \gamma(\Pi_{i=1}^n (1 - x_i)^{\delta_i})$$

Here δ_i denotes the relative importance degree of evaluation item value x_i.

When we use a compensatory operator Φ, the fuzzy integral has the following form:

$$\oint_A h(x) \circ g(\cdot) = \sup_{E \subseteq A} \{\min\{\Phi_{x \in E} h(x), g(E)\}\}$$

On the other hand, in the real world, it is possible for evaluation values to have uncertainty. Thus it is helpful to consider certainty factor or reliability of evaluation values in fuzzy integral. For this purpose, we use the interest focus A as a fuzzy set and interpret membership degrees as certainty factors. When we consider certainty factor, fuzzy integral has the following form:

$$\oint_A h(x) \circ g(\cdot) = \sup_{E \subseteq Supp(A)} \{\min\{\min_{x \in E} Q(\mu_A(x), h(x)), \, g(E)\}\}$$

In the formula, $Supp(A)$ denotes the support[11] of fuzzy set A, and $Q(\mu_A(x), h(x))$ indicates the application of certainty factor $\mu_A(x)$ to evaluation value $h(x)$. For Q operation, we can use either minimum or product operations. When a compensatory operator is used in fuzzy integral, the above formula comes to be as follows:

$$\oint_A h(x) \circ g(\cdot) = \sup_{E \subseteq Supp(A)} \{\min\{\Phi_{x \in E} Q(\mu_A(x), h(x)), \, g(E)\}\}$$

4   Information Aggregating Network

When we extend fuzzy integral by introducing compensatory operators, we should apply compensatory operator to all elements of the power set of evaluation item set X. Thus there is computational burden in performing fuzzy integral. In addition, sometimes we should determine compensation degree γ and/or relative importance degrees δ_i. It is reasonable to determine such parameters from human-provided data by learning. To tackle these problems, we

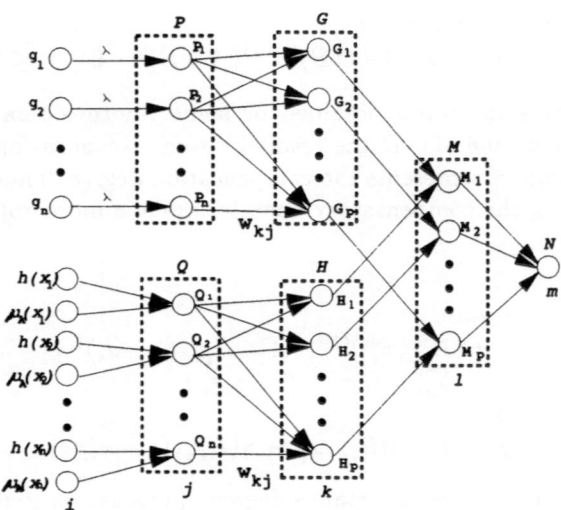

Fig. 2. Proposed network model

propose a network model that carries out fuzzy integral by parallel computations and provides a mechanism for tuning parameters such as γ, δ_i.

Figure 2 shows the proposed network model which performs the extended fuzzy integral and uses λ-fuzzy measure values. In the network model, each node has its own role and the connection weights between nodes are fixed to 1 except the connections between layer P and its input layer which have λ value.

Let the evaluation item set $X = \{x_1, x_2, \ldots, x_n\}$. The layers P and Q play the role of evaluating λ-fuzzy measure values. The fuzzy density values g_is and λ value obtained by fuzzy measure identification method from human-provided measure value come to be input for the layer P. The layer P contains n nodes and the layer Q contains $2^n - 1$ nodes. A node of layer Q corresponds to a subset E of $X (E \subseteq X)$ and produces fuzzy measure value of E, i.e., $g(E) = [\Pi_{x_i \in E}(1 + \lambda g_i) - 1]/\lambda$. The node P_j corresponding to evaluation item x_j and the node G_k corresponding to a subset E of X, is connected when $x_j \in E$. For this connection, its weight w_{kj} is set to 1. The node operations of layer P and Q are as follows:

$$P_j = \lambda g_j$$
$$G_k = [\Pi_j w_{kj}(1 + P_j) - 1]/\lambda$$

The layer Q plays the role to restrict the domain of fuzzy integral to the interest focus A. When A is a crisp set, $\mu_A(x_i) = 1$ if $x_i \in A$ and $\mu_A(x_i) = 0$ otherwise. When A is a fuzzy set, $\mu_A(x_i)$ has a value on $[0, 1]$ and it indicates certainty factor. In the case of fuzzy set, x_is such that $\mu_A(x_i) > 0$ belong to the interest focus. The operation of layer Q is either minimum operation or product operation.

$$Q_j = \begin{cases} \min\{h(x_j), \mu_A(x_j)\} & \text{if min is used} \\ h(x_j) \cdot \mu_A(x_j) & \text{if product is used} \end{cases}$$

In the layer H, each node corresponds to a subset E of X and it has the role to carry out $\min_{x \in E} h(x)$ or $\Phi_{x \in E} h(x)$ where Φ denotes a compensatory operator. The connections between layer Q and H are established in the same way as the connections between layer P and G. The nodes of layer H perform the following operation.

$$H_k = \begin{cases} \min_j\{w_{kj}Q_j\} & \text{if min is used} \\ \Phi\{w_{kj}Q_j\} & \text{if } \Phi \text{ is used} \end{cases}$$

Each node of layer M performs the minimum operation to impose the restriction that the aggregated evaluation value can not be greater than the importance degree of the corresponding evaluation items. In the layers H and G, the nodes H_l and G_l correspond to the same evaluation items.

$$M_l = \min\{G_l, H_l\}$$

The node of layer N produces the overall evaluation value by sup operation.

$$N = \sup_l M_l$$

By these node operations, the proposed network model carries out both Sugeno's fuzzy integral and its extended one.

When we use a compensatory operator in fuzzy integral, the layer H itself has a parameter for the compensation degree γ, and each node of Q has a parameter for its relative importance degree δ_i. With training data, we can tune these parameters by the gradient descent method. Let the error E be the square sum of the differences between the real output N_i and the desired output d_i. The tuning of parameter γ is performed as follows:

$$E = \tfrac{1}{2} \sum_i (N_i - d_i)^2$$

$$\gamma(t+1) = \gamma(t) + \eta \frac{\partial E}{\partial \gamma}$$

$$\frac{\partial E}{\partial \gamma} = \sum_k \frac{\partial E}{\partial H_k} \frac{\partial H_k}{\partial \gamma}$$

$$\frac{\partial E}{\partial H_k} = \frac{\partial E}{\partial M_k} \frac{\partial M_k}{\partial H_k}$$

$$\frac{\partial M_k}{\partial H_k} = \begin{cases} 1 & \text{if } M_k = H_k \\ 0 & \text{otherwise} \end{cases}$$

When γ-model is used as a compensatory operator, the relative importance degrees δ_j of evaluation items are tuned as follows:

$$\delta_j(t+1) = \delta_j(t) + \eta \frac{\partial E}{\partial \delta_j}$$

$$\frac{\partial E}{\partial \delta_j} = \frac{\partial E}{\partial Q_j} \frac{\partial Q_j}{\partial \delta_j}$$

$$\frac{\partial E}{\partial Q_j} = \sum_k \frac{\partial E}{\partial H_k} \frac{\partial H_k}{\partial Q_j}$$

5 Experiment

To show the applicability of proposed method, we performed an experiment to determine preference degree of secondhand car with fuzzy integral, where evaluation items were mileage, appearance, expected maintenance cost, age of operation, and history of accidents.

To apply fuzzy integral, we first should determine the importance degrees for the evaluation items with the properties of fuzzy measures. Thus we intuitively assigned the human-provided measure values for the evaluation items as shown in Table 1. In Table 1, a row consisting of m, ap, c, ag, h represents a subset for the evaluation item set X, and the value of $g(\cdot)$ represents the importance degree for the evaluation items.

For these data, we applied the genetic algorithms for λ-fuzzy measure identification. Thus we obtained the following values: mileage 0.177170, appearance 0.302570, maintenance cost 0.309290, age of operation 0.233480, history of accident 0.350640, and $\lambda : -0.464951$.

With these identified values, we should apply fuzzy integral to the evaluation values for producing preference degrees.

Table 1. Data for importance degrees

m	ap	c	ag	h	$g(\cdot)$	m	ap	c	ag	h	$g(\cdot)$
1	0	0	0	0	0.2	0	1	0	0	0	0.3
0	0	1	0	0	0.3	0	0	0	1	0	0.25
0	0	0	0	1	0.35	1	1	0	0	0	0.45
1	0	1	0	0	0.45	1	0	0	1	0	0.4
1	0	0	0	1	0.5	0	1	1	0	0	0.6
0	1	0	1	0	0.55	0	1	0	0	1	0.6
0	0	1	1	0	0.5	0	0	1	0	1	0.65
0	0	0	1	1	0.5	1	1	1	0	0	0.7
1	1	0	1	0	0.55	1	1	0	0	1	0.75
1	0	1	1	0	0.56	1	0	1	0	1	0.74
1	0	0	1	1	0.65	0	1	1	1	0	0.78
0	1	1	0	1	0.83	0	1	0	1	1	0.73
0	0	1	1	1	0.81	1	1	1	1	0	0.87
1	1	1	0	1	0.93	1	1	0	1	1	0.89
1	0	1	1	1	0.89	0	1	1	1	1	0.95
1	1	1	1	1	1.0						

m: mileage, ap: appearance, c: maintenance cost,
ag: age of operation, h: history of accident

In the experiment we used the extended fuzzy integral, where the weighted arithmetic mean is used as a compensatory operator. Now we should determine the compensation parameter γ for the compensatory operator. Thus we constructed an information aggregating network proposed in the previous section. To tune the parameter, we gathered data consisting of evaluation values and preference degree as shown in Table 2. In Table 2, the values in the columns m, ap, c, ag, h

denote the evaluation values with respect to the corresponding evaluation items, and the values in the column v denotes the preference degree for the given evaluation values. Using the data in Table 2, we tuned the parameter by the method mentioned in the previous section. The tuning produced the γ value 0.15.

Through this procedure, we determine the fuzzy density values g_is and λ value, and the compensation parameter γ. Now we could produce the preference degree for the given evaluation values of evaluation items with the proposed network model. The produced results by the network showed similar to human-provided preference degrees.

Table 2. Data for parameter tuning

m	ap	c	ag	h	v	m	ap	c	ag	h	v
0.3	0.2	0.6	0.4	0.5	0.52	0.5	1.0	0.3	0.5	1.0	0.60
0.3	0.2	0.3	0.1	0.7	0.36	0.5	0.6	0.3	0.2	0.8	0.60
0.4	0.3	0.8	0.2	0.4	0.46	0.5	0.3	0.8	0.1	0.4	0.47
0.6	0.3	0.9	0.2	0.5	0.56	0.8	0.8	0.2	0.7	0.6	0.64
0.2	0.7	0.5	0.4	0.2	0.53	0.5	0.3	0.4	0.1	0.4	0.42
0.6	0.2	0.3	0.2	0.5	0.51	1.0	0.2	0.2	0.3	0.4	0.49
0.7	0.3	0.2	0.3	0.3	0.36	0.3	0.8	0.6	0.2	0.7	0.63

m: mileage, ap: appearance, c: maintenance cost,
ag: age of operation, h: history of accident

6 Conclusion

In this paper, we investigated the issues to be considered when we use Sugeno's fuzzy integral to aggregate information, and proposed solutions to deal with them. First, to identify importance degrees of evaluation items with the properties of fuzzy measures, we suggested the genetic algorithm-based λ-fuzzy measure identification method. Second, to improve the properties of fuzzy integral, we introduce compensatory operators into fuzzy integral. Thus we can avoid excessively emphasizing the pessimistic aspect in the information aggregation. Third, to lessen the computational burden in performing extended Sugeno's fuzzy integral, we proposed a network model to perform fuzzy integral. The proposed network model makes it possible to tune the parameters when a compensatory operator is used. In the experiment to determine preference degrees for second-hand car, we could see that the proposed method is useful.

The tuning method of the proposed network model is based on the gradient-descent method. Hence there is some possibility of falling into a local minima. As an alternative method for determining parameters of extended Sugeno's fuzzy integral, the genetic algorithm-based approach seems to be promising.

References

1. L. Davis, *Handbook of Genetic Algorithms*(eds.), Van Nostrand Reinhold:New York, 1991.

2. K.De Jong, Learning with Genetic Algorithms: An Overview, *Machine Learning*, Vol.3, pp.121-138, 1988.
3. J.M. Fitzpatrick, J.J. Grefenstette, Genetic Algorithms in Noisy Environments, *Machine Learning*, Vol.3, pp.101-120, 1988.
4. D.E. Goldberg, *Genetic Algorithms in Search, Optimization & Machine Learning*, Addison-Wesley, 1989.
5. K.-M. Lee, H. Leekwang, Genetic Algorithms for Fuzzy Measure Identification, *the 3rd International Conference on Fuzzy Logic, Neural Networks, and Soft Computing*(Iizuka, Japan), pp.461-463, 1994.
6. D. Ralescu, G. Adams, The Fuzzy Integral, *Journal of Mathematical Analysis and Applications*, Vol.75, pp.562-570, 1980.
7. M. Sugeno, Fuzzy Decision-Making Problems, *Trans. S.I.C.E.* Vol.11, No.6, pp.709-714, 1975.
8. H. Tahani, J. M. Keller, Information Fusion in Computer Vision using the Fuzzy Integral, *IEEE Systems, Man, and Cybernetics*, Vol.SMC-20, No.3, 1990.
9. S.T. Wierzchoń, An Algorithm for Identification of Fuzzy Measure, *Fuzzy Sets and Systems*, Vol.9, pp.69-78, 1983.
10. S.T. Wierzchoń, On Fuzzy Measure and Fuzzy Integral, *Fuzzy Information and Decision Processes*:M.M. Gupta and E. Sanchez(eds.), pp.79-86, 1982.
11. H.-J. Zimmermann, *Fuzzy Set Theory - and Its Applications*, Kluwer-Nijhoff Publishing: Boston, 364p, 1985.

A Neuro-Fuzzy Architecture for High Performance Classification*

Sung-Bae Cho

Department of Computer Science, Yonsei University
134 Shinchon-dong, Sudaemoon-ku, Seoul 120-749, Korea

Abstract. The concept of combining modular neural networks has been recently exploited as a new direction for the development of highly reliable neural network systems in the area of pattern classification. In this paper we present an efficient method for combining the modular networks based on fuzzy logic, especially the fuzzy integral. This method nonlinearly combines objective evidences, in the form of network outputs, with subjective evaluation of the reliability of the individual neural networks. Also, for more effective aggregation, we adopt the extension of the fuzzy integral with ordered weighted averaging operators. The experimental results with the recognition problem of on-line handwriting characters show that the performance of individual networks could be improved significantly.

1 Introduction

In the past several years, there has been a tremendous growth in the complexity of the recognition, estimation and control problems expected from neural networks. In solving these problems, we are faced with a large variety of learning algorithms and a vast selection of possible network architectures. After all the training, we choose the best network with a winner-takes-all cross-validatory model selection. However, recent theoretical and experimental work indicates that we can improve performance by considering methods for combining neural networks [1, 2, 3, 4, 5, 6]. One of the key issues of this approach is how to combine the results of the various networks to give the best estimate of the optimal result. There are a number of possible schemes for automatically optimizing the choice of individual networks and/or combining architectures.

A straight-forward approach is to decompose the problem into manageable ones for several different subnetworks and combine them via a gating network that decides which of the subnetworks should be used for each case. Hampshire and Waibel [7] have described a system of this kind that can be used when the decomposition into subtasks is known prior to training, and Jacobs *et al.*

---

* This work was partly conducted while at: KAIST Computer Science Department, 373-1 Koosung-dong, Yoosung-ku, Taejeon 305-701, Korea, sbchogorai.kaist.ac.kr; and ATR Human Information Processing Research Laboratories, 2-2 Hikaridai, Seika-cho, Soraku-gun, Kyoto 619-02, Japan, sbchohip.atr.co.jp

[3] have also proposed a supervised learning procedure for systems composed of many separate networks, each of which learns to handle a subset of the complete set of training instances. The subnetworks are local in the sense that the weights in one expert are decoupled from the weights in other subnetworks. However, there is still some indirect coupling because if some other network changes its weights, it may cause the gating network to alter the responsibilities that get assigned to the subnetworks.

An alternative one is to independently generate a number of networks for possible generalizers and utilize all of them for obtaining robust output. While a usual scheme chooses one best network from amongst the set of candidate networks based on a winner-takes-all strategy, this approach keeps multiple networks and runs them all with an appropriate collective decision strategy. This is different from the aforementioned "adaptive mixtures of local experts" [3], in the sense that here networks do not decompose the task but learn globally the same task with different points of view. Several methods for combining evidence produced by multiple information sources have been applied in statistics, management sciences, and pattern recognition [8, 9]. A general result from the previous works is that averaging separate networks improves generalization performance for the mean squared error [6]. If we have networks of different accuracy, however, it is obviously not good to take their simple average or simple voting.

To give a solution to the problem, this paper presents a fusion method that considers the difference of performance of each network in combining the networks, which is based on the notion of fuzzy logic, especially the fuzzy integral. This method combines the outputs of separate networks with importance of each network, which is subjectively assigned as the nature of fuzzy logic. Also, we demonstrate the superior performance of the presented method and compare with conventional averaging methods by thorough experiments. Although a serious theoretical investigation is beyond the scope of this paper, we will demonstrate the effectiveness of the method by experimental results on a difficult OCR problem. For more details, refer the forthcoming publication made by the author [10].

The rest of this paper is organized as follows. Section 2 formulates the modular neural networks and considers possible methods for combining them. In section 3, we present the proposed architecture combining the modular neural networks with the fuzzy integral, and extends it with ordered weighted averaging (OWA) operators. Shown in section 4 are results with the recognition of on-line handwriting characters.

2 Backgrounds

2.1 Modular Neural Networks

In this section, we present the modular neural network (MNN) which combines a population of neural network outputs to estimate a function $f(x)$ defined by

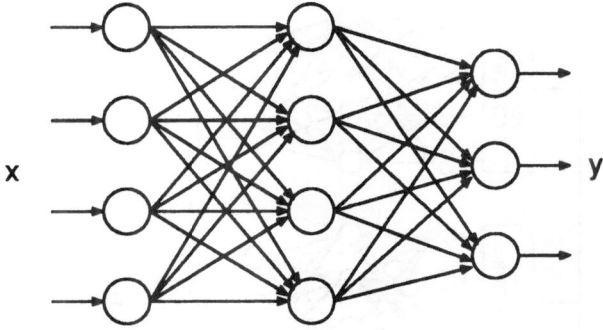

Fig. 1. A two-layered neural network architecture.

$f(x) = E[y|x]^2$ [11]. Fig. 1 shows a two-layer neural network classifier with T neurons in the input layer, H neurons in the hidden layer, and c neurons in the output layer. Here, T is the number of features, c is the number of classes, and H is an appropriately selected number.[3] The network is fully connected between adjacent layers. The operation of this network can be thought of as a nonlinear decision-making process; Given an unknown input $X = (x_1, x_2, \ldots, x_T)$ and the class set $\Omega = \{\omega_1, \omega_2, \ldots, \omega_c\}$, each output neuron estimates the probability $P(\omega_i|X)$ of belonging to this class by

$$P(\omega_i|X) \approx f\left\{\sum_{k=1}^{H} w_{ik}^{om} f\left(\sum_{j=1}^{T} w_{kj}^{mi} x_j\right)\right\}, \tag{1}$$

where w_{kj}^{mi} is a weight between the jth input neuron and the kth hidden neuron, w_{ik}^{m} is a weight from the kth hidden neuron to the ith class output, and f is a sigmoid function such as $f(x) = 1/(1 + e^{-x})$. The neuron having the maximum value is selected as the corresponding class.

This kind of network trains on a set of example patterns and discovers relationships that distinguish the patterns. A network of a finite size, however, does not often load a particular mapping completely or it generalizes poorly. Increasing the size and number of hidden layers most often does not lead to any improvements. Furthermore, in complex problems such as character recognition, both the number of available features and the number of classes are large. The features are neither statistically independent nor unimodally distributed.

---

[2] The outputs of neural networks are not just likelihoods or binary logical values near zero or one. Instead, they are estimates of Bayesian *a posteriori* probabilities of a classifier.

[3] There has been a long debate as to how to determine H as appropriate for any given problem. This has motivated the development of several constructive training techniques, such as Fahlman's *Cascade Correlation*.

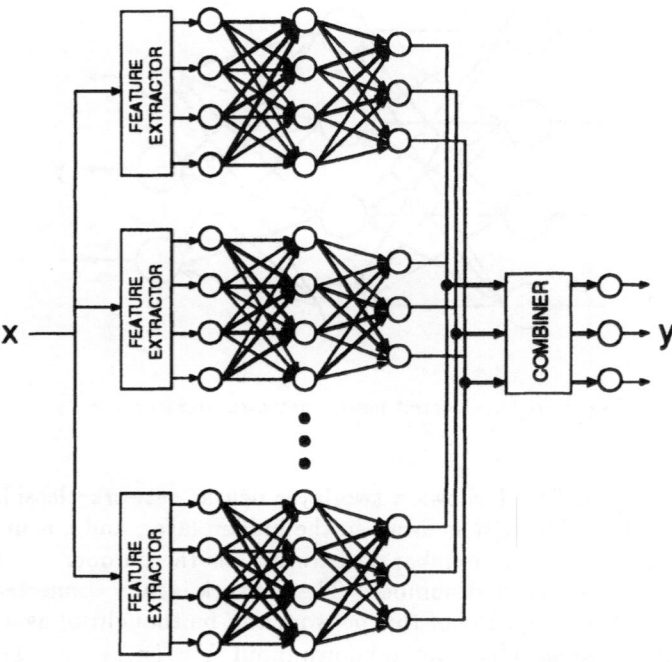

Fig. 2. The modular neural network scheme combined by fusion method.

The basic idea of the presented network scheme is to develop n independently trained neural networks with relevant features, and to classify a given input pattern by utilizing combination methods to decide the collective classification [1, 12] (see Fig. 2). Then it naturally raises the question of obtaining a consensus on the results of each individual network or expert.

2.2 Combining Methods

There might be two general approaches to combining the modular neural networks: one is based on fusion technique and the other on voting technique. In the methods based on the fusion technique, the classification of an input X is actually based on a set of real value measurements:

$$P(\omega_i|X), \quad 1 \leq i \leq c.$$

They represent the probabilities that X comes from each of the c classes under the condition X. In the modular network scheme, each network k estimates by itself a set approximations of those true values as follows:

$$P_k(\omega_i|X), \quad 1 \leq i \leq c, \ 1 \leq k \leq n.$$

One simple approach to combine the results on the same X by all n networks is to use the following average value as a new estimation of combined network:

$$P(\omega_i|X) = \frac{1}{n} \sum_{k=1}^{n} P_k(\omega_i|X), \quad 1 \le i \le c. \tag{2}$$

We can think of such a combined value as an averaged Bayes classifier. This estimation will be improved if we give the judge the ability to bias the outputs based on *a priori* knowledge about the reliability of the networks:

$$P(\omega_i|X) = \sum_{k=1}^{n} r_k P_k(\omega_i|X), \quad 1 \le i \le c, \tag{3}$$

$$\text{where } \sum_{k=1}^{n} r_k = 1. \tag{4}$$

Another alternative is to use the maximum value of $P_k(\omega_i|X)$ denoted by $P_m(\omega_i|X)$, to replace the correspondent average value. Since $\sum_{i=1}^{c} P_m(\omega_i|X) \ne 1$, we use the following normalized values as the new estimations:

$$P(\omega_i|X) = \frac{P_m(\omega_i|X)}{\sum_{j=1}^{c} P_m(\omega_j|X)}, \quad 1 \le i \le c. \tag{5}$$

The other method based on voting techniques considers the result of each network as an expert judgement. A variety of voting procedures can be adopted from group decision making theory: unanimity, majority, plurality, Borda count, and so on. In particular, we will introduce the two of them: majority voting and Borda count.

The majority voting rule chooses the classification made by more than half the networks. When there is no agreement among more than half the networks, the result is considered an error. To appreciate the network performance, let's assume that all neural networks arrive at the correct classification with a certain likelihood $1 - p$ and that they make independent errors. The chances of seeing exactly k errors among n copies of the network is then

$$\binom{n}{k} p^k (1-p)^{n-k} \tag{6}$$

which gives the following likelihood of the majority rule being in error

$$\sum_{k>n/2}^{n} \binom{n}{k} p^k (1-p)^{n-k}. \tag{7}$$

It can be shown by induction for odd n (or separately for even n) that provided $p < 1/2$, (7) is monotonically decreasing in n. In other words, if each network can get the correct answer more than half the time, and if network responses are independent, then the more networks used, the less the likelihood of an error by a majority decision rule. In the limit of infinite n, the coordinated error rate goes to zero.

For any particular class c, the Borda count is the sum of the number of classes ranked below c by each network; Let $B_j(c)$ be the number of classes ranked below the class c by the jth network. Then, the Borda count for class c is $B(c) = \sum_{j=1}^{n} B_j(c)$. The final decision is given by selecting the class label whose Borda count is the largest.

3 Neuro-Fuzzy Architecture

In this section, we shall describe the neuro-fuzzy architecture that utilizes the fuzzy integral for combining the modular neural networks. This method might produce better classification results, especially when we can assign the importance to each network.

3.1 Overview of Fuzzy Integral

The fuzzy integral is a nonlinear functional that is defined with respect to a fuzzy measure, especially g_λ-fuzzy measure introduced by Sugeno [13]. The ability of the fuzzy integral to combine the results of multiple sources of information has been established in several previous works [14, 15, 16]. In the following we shall introduce some definitions of it and present an effective method for combining the outputs of multiple networks with regard to subjectively defined importances of individual networks.

Definition 1. A set function $g : 2^X \to [0, 1]$ is called a fuzzy measure if
1) $g(\emptyset) = 0, g(X) = 1,$
2) $g(A) \le g(B)$ if $A \subset B$,
3) If $\{A_i\}_{i=1}^{\infty}$ is an increasing sequence of measurable sets, then

$$\lim_{i \to \infty} g(A_i) = g(\lim_{i \to \infty} A_i).$$

Note that g is not necessarily additive. This property of monotonicity is substituted for the additivity property of the measure.

From the definition of a fuzzy measure g, Sugeno introduced the so-called g_λ-fuzzy measures satisfying the following additional property: For all $A, B \subset X$ and $A \cap B = \emptyset$,

$$g(A \cup B) = g(A) + g(B) + \lambda g(A)g(B), \quad \text{for some } \lambda > -1.$$

It affords that the measure of the union of two disjoint subsets can be directly computed from the component measures.

Example 1. Consider the following case of $Y = \{y_1, y_2, y_3\}$ together with density values $g^1 = 0.34$, $g^2 = 0.32$, and $g^3 = 0.33$. Using the equation 11 (which will be introduced below), the Sugeno measure g must have a parameter λ satisfying $0.0359\lambda^2 + 0.3266\lambda - 0.001 = 0$. The unique root greater than -1 for this equation is $\lambda = 0.0305$, which produces the following fuzzy measure on the power set of Y:

Subset A	$g_{0.0305}(A)$
$\emptyset$	0
$\{y_1\}$	0.34
$\{y_2\}$	0.32
$\{y_3\}$	0.33
$\{y_1, y_2\}$	0.6633
$\{y_2, y_3\}$	0.6532
$\{y_1, y_3\}$	0.6734
$\{y_1, y_2, y_3\}$	1.0

As expected, the subset of criteria $\{y_1, y_3\}$ is more important for confirming the hypothesis than either subsets $\{y_1, y_2\}$ or $\{y_2, y_3\}$.

Using the notion of fuzzy measures, Sugeno developed the concept of the fuzzy integral, which is a nonlinear functional that is defined with respect to a fuzzy measure, especially g_λ-fuzzy measure [13, 14, 15].

Definition 2. Let X be a finite set, and $h : X \to [0, 1]$ be a fuzzy subset of X. The fuzzy integral over X of the function h with respect to a fuzzy measure g is defined by

$$h(x) \circ g(\cdot) = \max_{E \subseteq X} \left[\min \left(\min_{x \in E} h(x), g(E) \right) \right]$$
$$= \sup_{\alpha \in [0,1]} [\min (\alpha, g(h_\alpha))] \tag{8}$$

where h_α is the α level set of h,

$$h_\alpha = \{x \mid h(x) \geq \alpha\}. \tag{9}$$

The following properties of the fuzzy integral can be easily proved [15].

1) If $h(x) = c$, for all $x \in X$, $0 \leq c \leq 1$, then

$$h(x) \circ g(\cdot) = c.$$

2) If $h_1(x) \leq h_2(x)$ for all $x \in X$, then

$$h_1(x) \circ g(\cdot) \leq h_2(x) \circ g(\cdot).$$

3) If $\{A_i \mid i = 1, \ldots, n\}$ is a partition of the set X, then

$$h(x) \circ g(\cdot) \geq \max_{i=1}^{n} e_i,$$

where e_i is the fuzzy integral of h with respect to g over A_i. For further details on the properties of the fuzzy integral and associated fuzzy measures for aggregating information, see the recent publication made by Yager [16].

To get some intuition for the fuzzy integral we consider the following interpretation. $h(y)$ measures the degree to which the concept h is satisfied by y. The term $\min_{y \in E} h(y)$ measures the degree to which the concept h is satisfied by all the elements in E. Moreover, the value $g(E)$ is a measure of the degree to which the subset of objects E satisfies the concept measured by g. Then, the value obtained from comparing these two quantities in terms of the min operator indicates the degree to which E satisfies both the criteria of the measure g and $\min_{y \in E} h(y)$. Finally, the max operation takes the biggest of these terms. One can interpret the fuzzy integral as finding the maximal grade of agreement between the objective evidence and expectation.

3.2 Fuzzy Integral for Network Fusion

The calculation of the fuzzy integral when Y is a finite set is easily given. Let $Y = \{y_1, y_2, \ldots, y_n\}$ be a finite set and let $h : Y \to [0, 1]$ be a function. Suppose $h(y_1) \geq h(y_2) \geq \ldots \geq h(y_n)$), (if not, Y is rearranged so that this relation holds). Then a fuzzy integral, e, with respect to a fuzzy measure g over Y can be computed by

$$e = \max_{i=1}^{n} [\min (h(y_i), g(A_i))] \tag{10}$$

where $A_i = \{y_1, y_2, \ldots, y_i\}$.

Note that when g is a g_λ-fuzzy measure, the values of $g(A_i)$ can be determined recursively as

$$g(A_1) = g(\{y_1\}) = g^1$$
$$g(A_i) = g^i + g(A_{i-1}) + \lambda g^i g(A_{i-1}), \quad \text{for } 1 < i \leq n.$$

λ is given by solving the equation

$$\lambda + 1 = \prod_{i=1}^{n}(1 + \lambda g^i) \tag{11}$$

where $\lambda \in (-1, +\infty)$, and $\lambda \neq 0$. This can be easily calculated by solving an $(n-1)$st degree polynomial and finding the unique root greater than -1. Thus the calculation of the fuzzy integral with respect to a g_λ-fuzzy measure would only require the knowledge of the density function, where ith density, g^i, is interpreted as the degree of importance of the source y_i towards the final evaluation.

Let $\Omega = \{\omega_1, \omega_2, \ldots, \omega_c\}$ be a set of classes of interest. Note that each ω_i may, in fact, be a set of classes by itself. Let $Y = \{y_1, y_2, \ldots, y_n\}$ be a set of neural networks, and A be the object under consideration for recognition. Let $h_k : Y \to [0, 1]$ be the partial evaluation of the object A for class ω_k, that is, $h_k(y_i)$ is an indication of how certain we are in the classification of object A to be in class ω_k using the network y_i, where a 1 indicates absolute certainty that

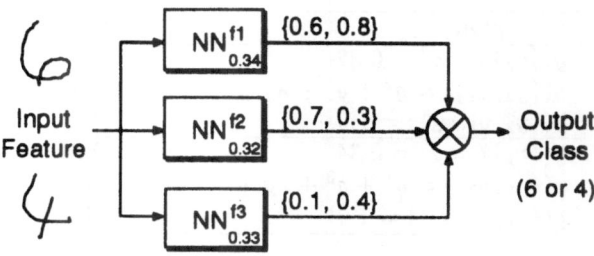

Fig. 3. A simple example of network outputs for a two class problem.

the object A is really in class ω_k and 0 implies absolute certainty that the object A is not in ω_k.

Corresponding to each y_i the degree of importance, g^i, of how important y_i is in the recognition of the class ω_k must be given. These densities can be subjectively assigned by an expert, or can be induced from data set. The g^i's define the fuzzy density mapping. Hence λ is calculated using (11) and thereby the g_λ-fuzzy measure, g, is constructed. Now, using (10) to (11), the fuzzy integral can be calculated. Finally, the class ω_k with the largest integral value is chosen as the output class. The following algorithm summarizes how network integration is performed with the fuzzy integral.

Algorithm : Network fusion by fuzzy integral
  calculate λ; /* importance of each net */
  for each class ω_k **do**
      for each neural network y_i **do**
          calculate $h_k(y_i)$;
          determine $g_k(\{y_i\})$;
      end_for
      compute the fuzzy integral;
  end_for
  determine the final class;

Example 2. Using the Example 1, how the consensus decision is performed by the fuzzy integral can now be described for a two class problem, which discriminates handwriting characters 6 and 4. Suppose that we obtain the network outputs for an input image as shown in Fig 3: $h(y_1) = 0.6$, $h(y_2) = 0.7$, and $h(y_3) = 0.1$, for class 1. For class 2, $h(y_1) = 0.8$, $h(y_2) = 0.3$, and $h(y_3) = 0.4$. The following table shows how the consensus is formed, where $H(E) = min(h(y_i), g(A_i))$.

Class	$h(y_i)$	$g(A_i)$	$H(E)$	$max[H(E)]$
1	0.7	$g(\{y_2\}) = g^2 = 0.32$	0.32	
	0.6	$g(\{y_2, y_1\}) = g^2 + g^1 + \lambda g^2 g^1 = 0.66$	0.6	✓
	0.1	$g(\{y_2, y_1, y_3\}) = 1.0$	0.1	
2	0.8	$g(\{y_1\}) = g^1 = 0.34$	0.34	
	0.4	$g(\{y_1, y_3\}) = g^1 + g^3 + \lambda g^1 g^3 = 0.67$	0.4	✓
	0.3	$g(\{y_1, y_3, y_2\}) = 1.0$	0.3	

Finally, the class 1 is selected as output. In the meantime, in case that we use the weighted average instead of the fuzzy integral, the class 2 is chosen as the correct class because the class 2 yields 0.5 ($0.34 \times 0.8 + 0.32 \times 0.3 + 0.33 \times 0.4$) whereas the class 1 produces 0.46 ($0.34 \times 0.6 + 0.32 \times 0.7 + 0.33 \times 0.1$). This example shows how the minute differences of g^i conspire so as to dramatically change the performance compared to simple averaging.

3.3 Extension of Fuzzy Integral with OWA Operators

In [16] Yager extended the fuzzy integral with two special families of OWA operators, S-OWA-AND and S-OWA-OR.

Definition 3. A mapping F from

$$I^n \rightarrow I (\text{where } I = [0, 1])$$

is called an OWA operator of dimension n if associated with F is a weighting vector W,

$$W = \begin{bmatrix} W_1 \\ W_2 \\ \vdots \\ W_n \end{bmatrix}$$

such that
1) $W_i \in (0, 1)$
2) $\sum_i W_i = 1$
and where

$$F(a_1, a_2, \cdots, a_n) = W_1 b_1 + W_2 b_2 + \cdots + W_n b_n,$$

where b_i is the ith largest element in the collection $a_1, a_2, \cdots, a_n$.

In [17] Yager shows how different assignment of the weights allows implementation of different quantifiers. For example, W^*, with $W_1 = 1$ and $W_i = 0$, $i \neq 1$ provides the max operator. W_* with $W_n = 1$ and $W_i = 0$ for $i \neq n$ gives us the min operator. In addition, $W_i = 1/n$ gives us the average $1/n \sum a_i$. This shows the more of the weights near the bottom the more "and-like" the aggregation while the more of weights near the top the more "or-like" the aggregation.

There are two special families of OWA operators which are useful for extending the fuzzy integral [16]. These are called the S-OWA-AND and S-OWA-OR operators. The S-OWA-AND operators are defined such that

$$\hat{F}_\alpha(a_1, \cdots, a_n) = \frac{1-\alpha}{n} \sum_i a_i + \alpha \min_i a_i.$$

The S-OWA-AND operators provide for *and-like* aggregations. In the formulation for the fuzzy integral we can obtain the effect of S-OWA-AND operators by replacing $\min_{x \in E} h(x)$ with

$$\frac{1-\alpha}{\text{Card} E} \sum_{x \in E} h(x) + \alpha \min_{x \in E} h(x).$$

The parameter α lies in the unit interval. The closer α is to one the more it becomes *and-like* aggregation.

On the contrary, S-OWA-OR operator provides for an *or-like* aggregation. This operator is defined such that

$$\tilde{F}_\beta(a_1, \cdots, a_n) = \frac{1-\beta}{n} \sum_i a_i + \beta \max_i a_i.$$

This provides for an *or-like* aggregation. Here again the parameter β lies in the unit interval and the closer β is to 1, the more like a pure *or* the operation. This S-OWA-OR operator can be used to provide a further generalization of the fuzzy integral. Let us denote $\min(\min_{x \in E} h(x), g(E))$ as $H(E)$. The value of the fuzzy integral is requiring that at *least one* subset E of X satisfy $H(E)$. With n the cardinality of X we can change the aggregation to

$$\frac{1-\beta}{2^n} \sum_{E \subset X} H(E) + \beta \max_{E \subset X} H(E).$$

With this change, depending on the choice of β, we are requiring that *some* or *few* of the E satisfy $H(E)$ rather than just one.

4 Experimental Results

In order to give an idea of practical application of the presented method to pattern recognition, a data set of handwriting characters was used as a source of both training and test samples. Handwriting characters were inputed to the computer (SUN workstation) by a Photron FIOS-6440 LCD tablet, which samples at the rate of 80 dots per second. The tasks were to classify Arabic numerals, uppercase letters, and lowercase letters collected from 13 different writers. The writers were told to draw the numerals and letters into prepared square boxes in order to facilitate segmentation.

An input character consists of a set of strokes, each of which begins with a pen-down movement and ends with a pen-up movement. Several preprocessing algorithms were applied to successive data points within each stroke to reduce quantization noise and fluctuations due to the writer's pen motion. The processes used were wild point reduction, dot reduction, hook analysis, three point smoothing, peak preserving filtering, and N point normalization [18]. Data points, representing single characters, were resampled with a fixed number of regularly spaced points. Then, a sequence of preprocessed data points was approximated by a sequence of 8-directional straight-line segments—the chain code, as used by Freeman [19].

To evaluate the performance of the proposed method, we implemented three different networks, each of which is a two-layered neural network having a different number of input neurons and 20 hidden neurons. NN_1, NN_2 and NN_3 have 10, 15, and 20 input neurons, respectively. In each case, the network makes a decision based on its resolution. For example, NN_1 uses sparsely sampled inputs, and in doing so is able to overcome variations in input noise. NN_3, on comparison, uses a finer view of the input image. The selection of the features is largely adhoc and no attempt was made to find an optimal coding scheme although this is an important issue in character recognition schemes. Our objective here is to evaluate and compare different fusion methods through an example which has a certain complexity and practical significance.

Each of the three networks was trained by the EBP algorithm with 40 samples per class, validated with another 500 samples, and tested on ten sets of additional samples collected from different 10 writers: The training process was stopped when the recognition rate over the validation set was optimized. This process and early stopping mechanism were adopted mainly for preventing networks from overtraining. The initial parameter values used for training were: Learning rate is 0.4 and momentum parameter is 0.6. An input vector is classified as belonging to the output class associated with the highest output activation. Each of the following experiments consisted of 10 trials in which the different data were made from different writers.

First, the behavior of the fuzzy integral of a function h with respect to a g_λ-fuzzy measure, g is examined. Table 1 shows these results. Here, each case shows a set of fuzzy densities corresponding to three networks and the recognition rates of numerals, uppercase letters, and lowercase letters using the fuzzy integral on the three networks. Using (11), the Sugeno measure g must have a parameter λ satisfying $0.006\lambda^2 + 0.11\lambda - 0.4 = 0$. The unique root greater than -1 for this equation is $\lambda = 3.109$.

As expected, the recognition results in the table depend on the g values. When the g values change, the new fuzzy integral value will change depending on how these changes are balanced with respect to the source corresponding to the fuzzy integral value. We assigned the fuzzy densities g^i, the degree of importance of each network, based on how good these networks performed on validation data. We computed these values as follows:

$$g^i = \frac{p_i}{\sum_j p_j} \cdot d\text{sum}, \qquad (12)$$

where p_i is the performance of network NN_i for the validation data and $d\text{sum}$

Table 1. The recognition rates of the fuzzy integral for different densities (%).

Case	g^1	g^2	g^3	Numeral	Upper	Lower
1	0.1	0.2	0.3	78.4	71.6	65.4
2	0.1	0.3	0.2	79.0	73.4	66.8
3	0.2	0.1	0.3	78.8	72.2	64.8
4	0.2	0.3	0.1	79.4	73.8	69.2
5	0.3	0.1	0.2	79.8	74.0	66.2
6	0.3	0.2	0.1	80.2	75.2	70.4

is the desired sum of fuzzy densities. The real values of these densities with the corresponding λ are shown in table 2.

Table 2. Fuzzy densities and the corresponding λ.

Subject	g^1	g^2	g^3	λ
Numerals	0.3430	0.3330	0.3230	−0.0149
Uppercases	0.3447	0.3312	0.3240	0.0003
Lowercases	0.3370	0.3321	0.3312	−0.0009

Table 3 reports the results of network fusion using the fuzzy integral on three different networks for numerals. In this table the value in the parentheses represent the confidence of the evaluation result. As can be seen, cases 2 and 3 were misclassified by NN_3 and NN_2, respectively. However, in the final evaluations they were correctly classified. In cases 5 and 17, one network with strong evidence overwhelmed the other networks, producing correct classification. Furthermore, in case 15, the fuzzy integral made a correct decision despite that the partial decisions from the individual neural networks were completely inconsistent. The effect of misclassification by the other networks has given rise to small fuzzy integral values for the correct classification in this case.

Table 4 shows the recognition rates of numerals, uppercase letters, and lowercase letters with respect to the three different networks and their combinations by utilizing consensus methods like majority voting, average, and the fuzzy integral. All results are averaged over ten different sets of the data. In this table, NN_1 to NN_3 represent the three individual networks, and NN_{all} a large network trained with all the features used by each network.

Although the network learned the training set almost perfectly in all three cases, the performances on the test sets are quite different. Furthermore, we can see that the performance did not improve by training a large network with considering all the features used by each network. This is a strong evidence that modular neural network might produce better result than conventional single network approach. The following test can further support to determine whether the fuzzy integral method is superior to the conventional method or not.

Table 3. Results of network fusion using the fuzzy integral on three different networks for numerals.

Data index	Actual class	Partial decision NN_1	Partial decision NN_2	Partial decision NN_3	Fuzzy integral decision
1	5	5 (0.9859)	5 (0.8995)	5 (0.9941)	5 (0.9598)
2	6	6 (0.9968)	6 (0.9985)	5 (0.3301)	6 (0.6877)
3	8	8 (0.9999)	0 (0.0022)	8 (0.9996)	8 (0.6668)
4	2	2 (0.9922)	2 (0.9998)	2 (0.9920)	2 (0.9946)
5	7	8 (0.0162)	8 (0.0087)	7 (0.9615)	7 (0.3205)
6	9	8 (0.0001)	7 (0.1195)	7 (0.4780)	7 (0.1991)
7	7	6 (0.0137)	7 (0.9903)	7 (0.9988)	7 (0.6630)
8	7	2 (0.1342)	7 (0.9677)	7 (0.0023)	7 (0.3233)
9	1	1 (0.9999)	1 (0.9972)	1 (0.9993)	1 (0.9988)
10	0	6 (0.7116)	6 (0.4098)	8 (0.8098)	6 (0.3753)
11	7	1 (0.1794)	8 (0.0003)	1 (0.0080)	1 (0.0625)
12	4	4 (0.9998)	4 (0.9999)	9 (0.9964)	4 (0.6694)
13	9	9 (0.9965)	9 (0.9958)	8 (0.0740)	9 (0.6691)
14	3	3 (0.9987)	3 (0.9912)	3 (0.9999)	3 (0.9966)
15	7	8 (0.0365)	0 (0.0460)	7 (0.4831)	7 (0.1610)
16	5	5 (0.9311)	3 (0.1304)	3 (0.6245)	5 (0.3265)
17	2	8 (0.3470)	2 (0.9983)	8 (0.2092)	2 (0.3327)
18	8	8 (0.9899)	0 (0.9669)	8 (0.6815)	8 (0.8384)
19	0	1 (0.0353)	4 (0.0004)	4 (0.0004)	1 (0.0118)
20	9	9 (0.7519)	9 (0.3799)	9 (0.9540)	9 (0.6953)
21	0	8 (0.8032)	0 (0.9994)	0 (0.9993)	0 (0.6662)
22	4	4 (0.9997)	9 (0.9170)	4 (0.9871)	4 (0.7099)
23	9	9 (0.9989)	1 (0.9944)	9 (0.9902)	9 (0.6705)
24	4	4 (0.9998)	4 (0.9999)	4 (0.9995)	4 (0.9997)
25	8	8 (0.9998)	0 (0.9882)	8 (0.8353)	8 (0.6204)

Table 4. Means and standard deviations of recognition rates (%).

Nets	Numeral Mean	Numeral S.D.	Uppercase Mean	Uppercase S.D.	Lowercase Mean	Lowercase S.D.
NN_1	82.6	6.36	73.2	8.95	73.9	7.73
NN_2	81.2	7.16	68.6	9.14	71.8	8.86
NN_3	81.0	7.15	70.8	10.60	72.1	9.30
NN_{all}	77.6	6.31	72.1	8.93	74.7	10.01
Voting	84.9	8.31	74.0	9.28	74.6	7.97
Average	86.9	7.24	75.2	9.95	78.2	8.85
Fuzzy	88.1	7.14	76.1	9.85	80.3	7.24

For a given test problem, let f_i^a denote the solution at convergence for method a using test data i. To test whether methods a and b have the same mean solution value, we compute the following statistic:

$$t = \frac{\sqrt{n}\bar{x}}{\sqrt{\frac{1}{n-1} \sum_{i=1}^{n}(x_i - \bar{x})^2}} \qquad (13)$$

where $n = 10$, $x_i = f_i^a - f_i^b$, and $\bar{x} = \frac{1}{n} \sum_{i=1}^{n} x_i$. (In this case the method b is of the fuzzy integral.) From this value we can reject the null hypothesis that $H_0 : \bar{x} \leq 0$ in favor of the alternative that $\bar{x} > 0$ with significance level α, where $\alpha = \Phi(t)$ and $\Phi(t)$ can be obtained from the table of percentage points of the t-distribution.

Since it follows t-distribution, an α point can be computed as the threshold t_α, where α could be 95, 99.95 or 99.9%, Then, if

$$|t| > t_\alpha \qquad (14)$$

the null hypothesis is rejected at a $100\% - \alpha$ level of significance, i.e., the fuzzy integral method is superior to the conventional method. Otherwise, the null hypothesis is accepted, i.e., we cannot say the fuzzy integral method improves the performance significantly.

Table 5 shows the results of the test with $n = 10$ for all three tasks. In this comparison, f_i^b is of the fuzzy integral, and f_i^a of the average method as mentioned in the equation (2) or the weighted average method in (3). These methods were chosen for comparisons because they had produced the best results among the conventional methods mentioned in this paper. In this comparison,

Table 5. t-test. (degree of freedom $= 9$, $t_{0.05} = 1.833$, $t_{0.025} = 2.262$, $t_{0.01} = 2.821$. "A" stands for Average method, and "WA" for Weighted Average. "Yes" indicates that the hypothesis is rejected for the task at the associated level of significance.)

Task	t	Significance Level		
		5 %	2.5 %	1 %
Numerals (A)	−2.908	Yes	Yes	Yes
Numerals (WA)	−2.575	Yes	Yes	No
Uppercase (A)	−2.193	Yes	No	No
Uppercase (WA)	−1.300	No	No	No
Lowercase (A)	−3.806	Yes	Yes	Yes
Lowercase (WA)	−3.361	Yes	Yes	Yes

the degree of freedom is $(n - 1) = 9$, and the threshold t_α with $\alpha = 95$, 97.25, and 99% is 1.833, 2.262 and 2.821, respectively. It is seen from table 5 that all the values of t, except for the weighted average of the uppercase letter task, are greater than t_α with $\alpha = 95\%$. Therefore, for all the cases except that, "no-improvement" hypothesis is rejected at a 5% level of significance. Similarly, other cases can be tested. This is a strong evidence that the proposed method is superior to the conventional methods.

Fig. 4 shows the recognition rates of the presented method with the OWA operators for the three tasks. The results indicate that the performance of the fuzzy integral might be enhanced if we select the α and β parameters appropriately.

5 Concluding Remarks

This paper has presented a neuro-fuzzy architecture that produces an improved performance on real-world classification problem, especially handwriting character recognition. One of the important advantages of the method is that not only is the classification results combined but that the relative importance of the different networks is also considered. The experimental results for classifying a large set of on-line handwriting characters show that it improves the generalization capability significantly. This indicates that even these straightforward, computationally tractable approach can significantly enhance pattern recognition.

Future efforts will concentrate on refining the feature extraction to capture more information, and testing the efficacy of this fuzzy neural system on larger data sets. The complementary nature of the neural network and the fuzzy logic lead us to believe that á further refined fuzzy neural system will significantly improve the state-of-the-art pattern recognizers, especially in noisy environments.

Acknowledgements

The author would like to thank Dr. K. Shimohara and Dr. Y. Tohkura at ATR HIP laboratories for continuous encouragement. This work was supported in part by a grant from the Korea Science and Engineering Foundation (KOSEF) and Center for Artificial Intelligence Research (CAIR), the Engineering Research Center (ERC) of Excellence Program.

References

1. Hansen, L.K., Salamon, P.: Neural network ensembles. IEEE Trans. Patt. Anal. Mach. Inte. **12** (1990) 993–1001
2. Scofield, C., Kenton, L., Chang, J.: Multiple neural net architectures for character recognition. Proc. Compcon. (1991) 487–491 (San Francisco, CA, IEEE Computer Society Press)
3. Jacobs, R.A., Jordan, M.I., Nowlan, S.J., Hinton, G.E.: Adaptive mixtures of local experts. Neural Comp. **3** (1991) 79–87
4. Wolpert, D.: Stacked generalization. Neural Net. **5** (1992) 241–259
5. Cho, S.-B., Kim, J.H.: Two design strategies of neural network for complex classification problems. Proc. 2nd Int. Conf. Fuzzy Logic & Neural Net. (1992) 759–762
6. Perrone, M.P., Cooper, L.N.: When networks disagree: Ensemble methods for hybrid neural networks. Neural Net. for Speech and Image Proc. (1993) (Mammone, R.J. ed., Chapman-Hill, London)

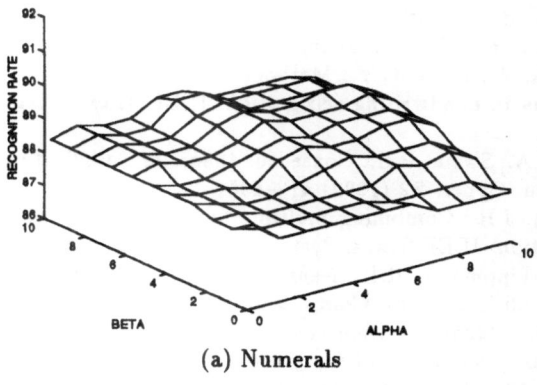

(a) Numerals

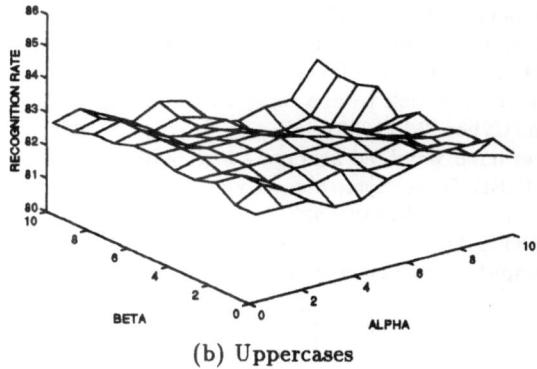

(b) Uppercases

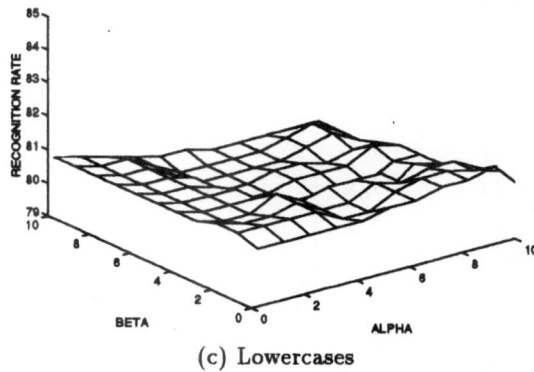

(c) Lowercases

Fig. 4. The recognition rates with OWA operators.

7. Hampshire II, J.B., Waibel, A.: Connectionist architectures for multi-speaker phoneme recognition. Adv. Neural Info. Proc. Syst **2** (1990) 203–210

8. Xu, L., Krzyzak, A., Suen, C.Y.: Methods of combining multiple classifiers and their applications to handwriting recognition. IEEE Trans. Syst. Man. Cyber. **22** (1992) 688–704

9. Benediktsson, J.A., Swain, P.H.: Consensus theoretic classification methods. IEEE Trans. Syst. Man. Cyber. **22** (1992) 418–435

10. Cho, S.-B., Kim, J.H.: Combining multiple neural networks by fuzzy integral for robust classification. IEEE Trans. Syst. Man. Cyber. **25** (1995) 380–384

11. Richard, M.D., Lippmann, R.P.: Neural network classifiers estimate Bayesian *a posteriori* probabilities. Neural Comp. **3** (1991) 461–483

12. Shlien, S.: Multiple binary decision tree classifiers. Patt. Recog. **23** (1990) 757–763

13. Sugeno, M.: Fuzzy measures and fuzzy integrals: A survey. Fuzzy Automata Dec. Proc. (1977) 89–102 (Amsterdam: North Holland)

14. Leszeynski, K., Penczek, P., Grochulskki, W.: Sugeno's fuzzy measures and fuzzy clustering. Fuzzy Sets Syst. **15** (1985) 147–158

15. Tahani, H., Keller, J.M.: Information fusion in computer vision using the fuzzy integral. IEEE Trans. Syst. Man. Cyber. **20** (1990) 733–741

16. Yager, R.R.: Element selection from a fuzzy subset using the fuzzy integral. IEEE Trans. Syst. Man. Cyber. **23** (1993) 467–477

17. Yager, R.R.: On ordered weighted averaging aggregation operators in multicriteria decisionmaking. IEEE Trans. Syst. Man. Cyber. **18** (1988) 183–190

18. Tappert, C.C., Suen, C.Y., Wakahara, T.: The state of the art in on-line handwriting recognition. IEEE Trans. Patt. Anal. Mach. Intel. **12** (1990) 787–808

19. Freeman, H.: Computer processing in line drawing images. Comp. Surv. **6** (1974) 57–98

Investigation of Stability and Robustness of a Fuzzy Traction Control System

Christian Schuster, Thomas Schmitz and Manfred Hiller

Department of Mechatronics
Gerhard-Mercator-University-GH-Duisburg
Tel. +49 203 3792199, Fax +49 203 3789143

Abstract In this paper the design of a fuzzy traction control system is presented. The parameters of the controller are determined with the help of the three dimensional vehicle dynamics simulation package FASIM. To investigate the robustness and stability of the controller, a large amount of parameters of the vehicle and the controller are varied. The results of the simulations are discussed in this paper.

1 Introduction

Since 1978, when the first anti-lock braking system has been developed, the significance of active vehicle components has continuously increased. One of the subsequent evaluations are traction control systems (ASR[1]). They are designed to limit the longitudinal slip at the driven wheels with the aim to improve the stability and traction of the vehicle, especially on inhomogenous or slippery road surfaces.

Modern controllers are increasingly pre-designed with the help of vehicle dynamics simulation programs. The development of the presented controller has been supported by the spatial vehicle dynamics simulation package FASIM. Due to the high complexity of the vehicle model (see section 3) it is nearly impossible to use common analytical methods of control theory for the design of active vehicle components. Therefore, in this paper, a new approach for the development of traction control systems using fuzzy logic is presented. The theoretical basis of fuzzy logic are only briefly discussed in this paper. A comprehensive overview can be taken from Zadeh 1973, Lee 1990, Driankov et al. 1993 and Zimmermann 1993.

---

[1] ASR: Antriebs–Schlupf–Regelung

2 Design of the fuzzy controller

There exist two different principles of operation of a traction control system. On the one hand the engine torque can be decreased by adjusting the servo throttle valve and by influencing the electronic ignition system and the fuel injection. On the other hand the surplus driving torque of each driven wheel can be reduced by controlled braking. Usually this will be performed by operating the solenoid valve at the brake cylinder of the corresponding wheel. These principles have to be applied depending on the velocity of the vehicle. Decreasing the engine torque with the throttle valve cannot be used at low velocities since stalling of the engine would be possible and braking cannot be used at high velocities to avoid wear (Table 1).

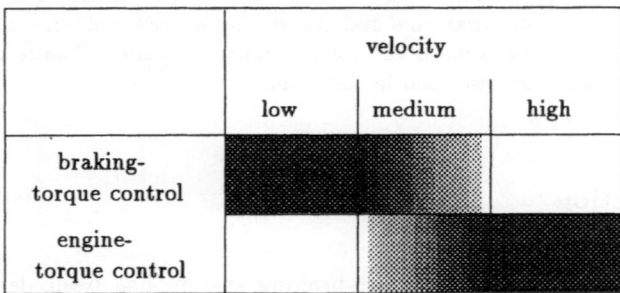

	velocity		
	low	medium	high
braking- torque control			
engine- torque control			

Table1.: Control concept depending on the vehicle velocity (see Schulze and Lissel 1986)

Besides the rotational speeds of the driven and free wheels ($\dot{\varphi}_{D_i}$, $\dot{\varphi}_{F_i}$, $i = 1, 2$) the driver input α_0 and the actual position of the accelerator pedal α are measured by sensors at discrete points of time (Figure 1).

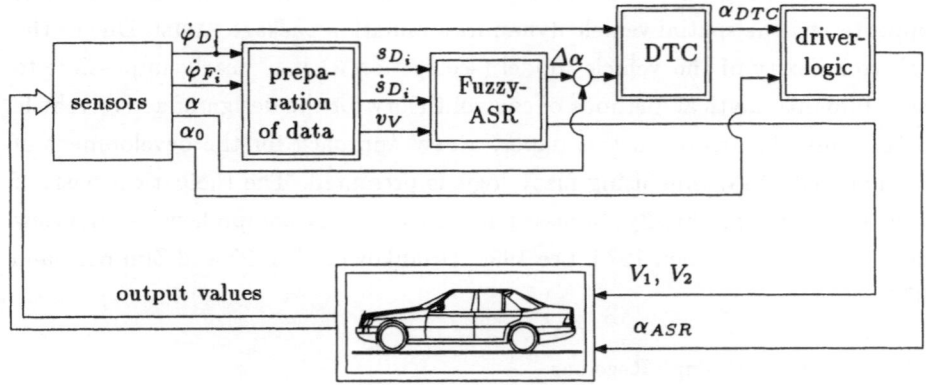

Figure1.: Block–diagram of the closed loop

With this data the input values of the fuzzy logic controller can be calculated. Since spinning of the free wheels is impossible, the vehicle velocity can be determined from the sensor signals for the rotational speeds of these wheels and the dynamic wheel radius r_R. The logitudinal slip of each driven wheel is given by

$$s_{D_i} = 1 - \frac{\dot{\varphi}_{F_i}}{\dot{\varphi}_{D_i}}, \quad i = 1, 2. \tag{1}$$

The slip–change is calculated as a difference quotient of two subsequent slip values:

$$\dot{s}_{D_i} \approx \frac{s_{D_i}(t) - s_{D_i}(t - \Delta t_R)}{\Delta t_R}, \quad i = 1, 2, \tag{2}$$

where Δt_R means the cycle time of the discrete controller. The change of the servo throttle valve $\Delta \alpha$ and the states of the solenoid valves at the driven wheels V_1, V_2 are the output values of the fuzzy logic controller which are given back to the vehicle.

Due to the verbal partitioning of the vehicle velocity in the linguistic values *low*, *medium* and *high* (see Table 1) it is suitable to design the braking–torque control only for *low* velocities and the engine–torque control only for *high* velocities. In the interval *medium* the two different control systems are fuzzy superposed.

The underlying membership functions for the inputs and outputs of the fuzzy logic controller are shown in Figure 2.

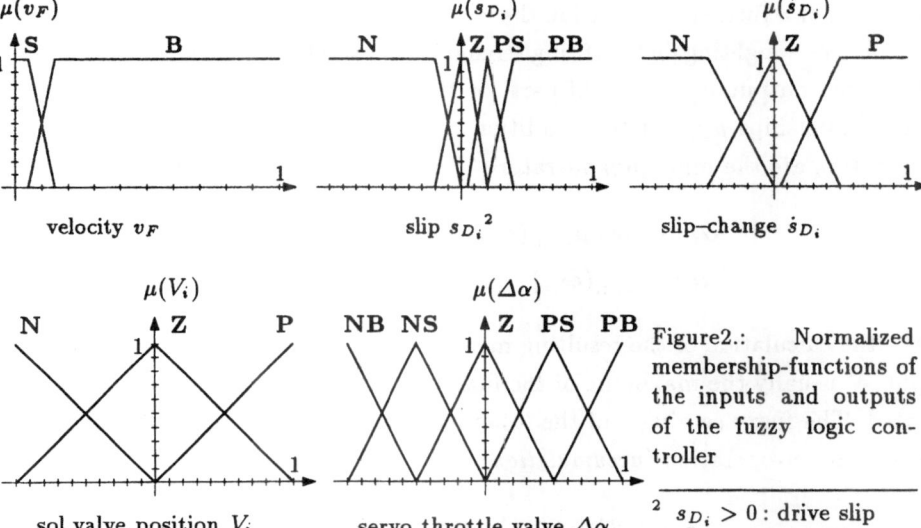

Figure2.: Normalized membership-functions of the inputs and outputs of the fuzzy logic controller

---

[2] $s_{D_i} > 0$: drive slip
$s_{D_i} < 0$: brake slip

For the values of the linguistic variables the common terms **Positive**, **Zero**, **Negative**, **Small** and **Big** are used. The braking torque of each driven wheel is controlled individually, i.e. the solenoid valves at the brake cylinders are only operated depending on the slip and the slip–change at the particular wheel. In contrast to this the engine–torque control operates on the select–low–principle. Here, the slip of the wheel with the worst friction conditions is responsible for the adjustment of the servo throttle valve. To reduce the number of necessary rules the formulation of the rulebase of the engine–torque control is initially generated separately for each driven wheel. In a second step the corresponding rulebases are superposed: the rule which causes the higher reduction of the servo throttle valve dominates the output of the engine–torque control. Table 2 shows the underlying rules by means of Karnaugh–Tables (see Bertram und Svaricek 1992).

s_{D_i}

$\dot{s}_{D_i}$		PB	PS	Z	N
	P	P	P	Z	N
	Z	P	P	N	N
	N	P	Z	N	N

s_{D_i}

$\dot{s}_{D_i}$		PB	PS	Z	N
	P	NB	NB	Z	Z
	Z	NB	NS	PB	NS
	N	NS	Z	PS	NB

a)   b)

Table 2.: Karnaugh-Tables of the rulebases for one driven wheel:
    a) braking–torque control,
    b) engine–torque control

In the following the decision-making logic of a fuzzy controller will be discussed briefly. For a further explanation the reader is referred to Lee 1990. The calculation of the weighting factor (firing strength) of a single rule is called *aggregation*. Here the output α_j of the 'if-part' of the rule is generated using the grades of memberships $\mu_{E_{j,i}}$ of the conditions of a single rule R_j. Usual aggregation operators are the minimum operator (3) and the algebraic product (4):

$$\alpha_j = \min\left(\mu_{E_{j,1}}\left(e_{0,1}\right), \dots, \mu_{E_{j,n}}\left(e_{0,n}\right) \right), \qquad (3)$$

$$\alpha_j = \mu_{E_{j,1}}\left(e_{0,1}\right) \cdot \mu_{E_{j,2}}\left(e_{0,2}\right) \cdots \mu_{E_{j,n}}\left(e_{0,n}\right). \qquad (4)$$

For the calculation of the resulting membership function $\mu_{A'}(a)$ of a fuzzy output A' usually the maximum of the fuzzy outputs A'_j of all fuzzy rules is calculated. The fuzzy sets $\mu_{A'_m}$ of the 'then-parts' of the rules are usually combined using the *or*-operation (*accumulation*)

$$\mu_{A'}(a) = \max\left(\mu_{A'_1}(a), \dots, \mu_{A'_m}(a) \right). \qquad (5)$$

The graphical interpretation of equations (3), (4) and (5) can be obtained from Figure 3. Here a controller with two inputs and one output is presented. The algebraic product is utilized as aggregation operator, in literature these techniques are known as *max-min-* and as *max-prod-methods*.

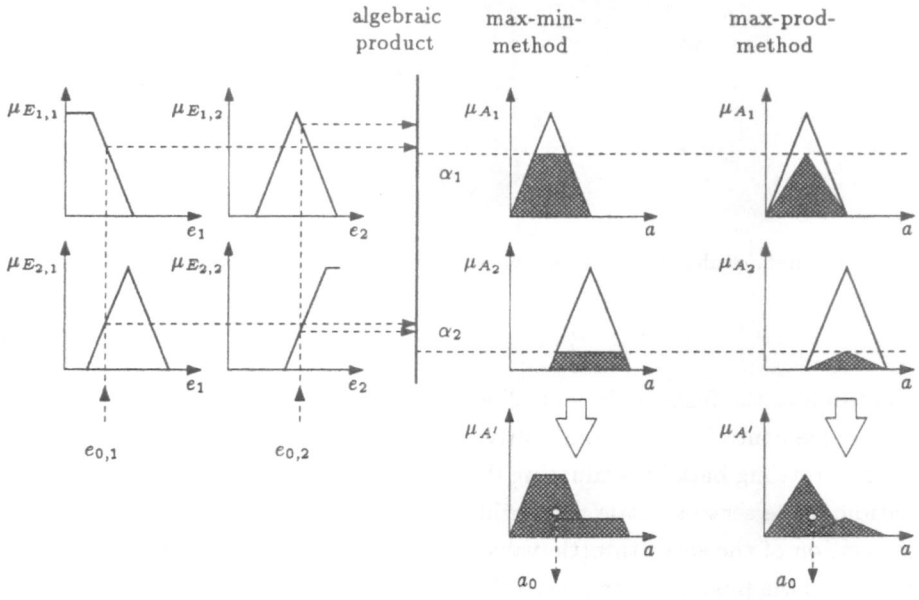

Figure3.: Mode of operation of a fuzzy controller

Figure 4 and Figure 5 present the control rules for a driven wheel as a normalized characteristic field. In this work the *max-prod-method* is utilized as inference mechanism for the fuzzy controller.

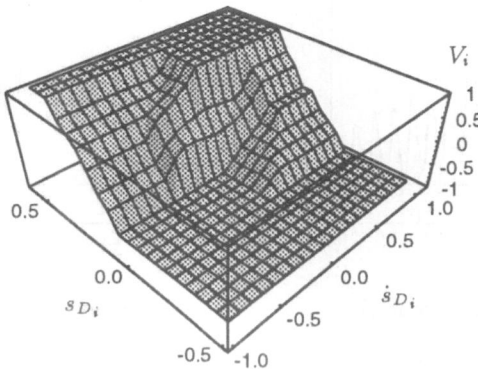

Figure4.: Rulebase of the braking-torque control as characteristic field

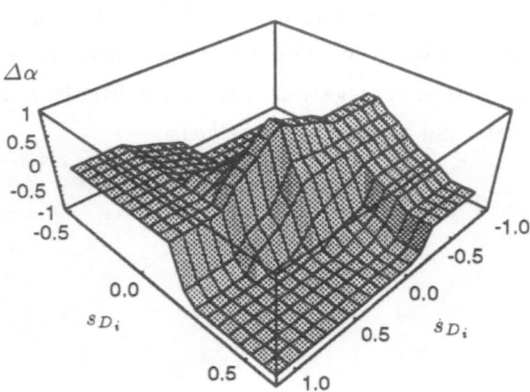

Figure5.: Rulebase of the engine–torque control as characteristic field

In addition to the fuzzy logic controller the traction control system consists of a drag torque control (DTC) and a driver logic. The DTC avoids drag torque in case of throttling back by evaluating the stationary motor field. If necessary, the position of the servo throttle valve will be increased. The driver logic compares the position of the servo throttle valve given by the ASR with the driver input and adapts the position of the servo throttle valve if necessary. The global mode of operation of the drag torque control can be taken from Figure 6 where the use of the engine torque control is demonstrated with and without DTC. Here a step steering manoeuvre at a high velocity is presented where wheelslip occurs and only the engine torque control is active.

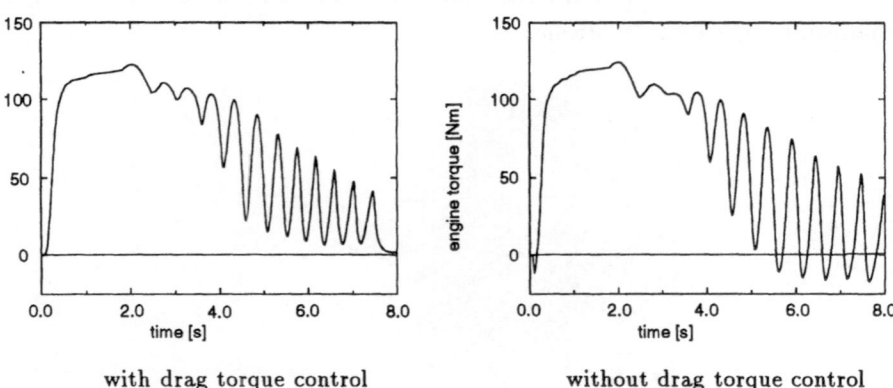

with drag torque control   without drag torque control

Figure6.: Mode of operation of the drag torque control

3   Modelling of the vehicle

The development costs of active vehicle components like anti–lock braking systems, traction control systems and active suspensions can be considerably decreased by using modern simulation techniques. FASIM is a modular simulation package for three-dimensional vehicle dynamics simulation including mechatronic elements like for example electronic controllers. A model library provides mathematical models for particular subsystems with different topologies, e.g. the library of the front suspensions contains the modules McPherson strut suspension, double wishbone strut suspension, etc. Thus, the user can build up a dynamical model of the vehicle under investigation quickly and easily. For the subsequent investigation two vehicle models with different concepts of power trains will be used (Figure 7).

a.) Mercedes Benz S300

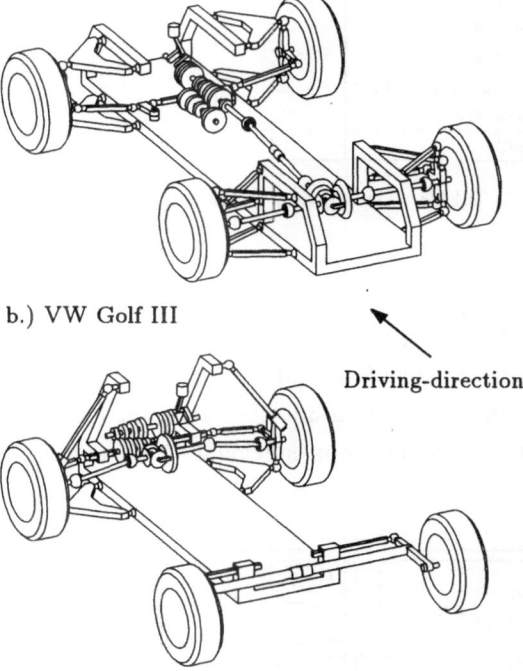

b.) VW Golf III

Driving-direction

Figure7.: Models of rear-driven and front driven vehicles

The Mercedes Benz S300 is a rear wheel driven car with a double wishbone front suspension and a five–link rear suspension. In opposite the VW Golf is a front wheel driven car with a McPherson front suspension and a trailing-arm torsion-beam rear suspension. The kinematical structure of both vehicles is given in Table 3. For further information concerning the modelling of the vehicles the reader should refer to Schnelle 1990 and Schmitz 1993.

subsystem	Mercedes Benz W140		
	degrees of freedom	bodies	kinematical loops
vehicle body	6	1	0
front suspension	3	11	5
rear suspension	2	12	10
engine suspension	2	1	0
power train	6	12	3
vehicle	19	37	18

subsystem	VW Golf		
	degrees of freedom	bodies	kinematical loops
vehicle body	6	1	0
front suspension	3	9	4
rear suspension	2	4	1
engine suspension	2	1	0
power train	5	10	2
vehicle	18	25	7

Table3.: Description of the vehicle models

The engine model used in this investigation is shown in Figure 8. It represents the superposition of a stationary model of the engine $T = T(\alpha, n)$ (e.g. calculated from the characteristic field of the engine), the engine torque T depending on the position of the throttle valve α and the rotational speed of the engine $\dot{\varphi}$, and $P_{T1}T_T$–elements to incorporate non–stationary processes.

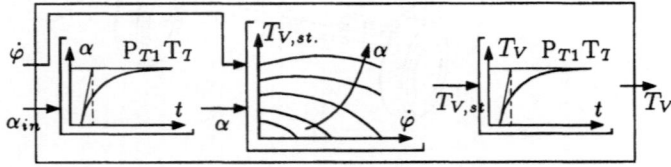

Figure8.: Simplified model of the engine

The dynamical behaviour of the hydraulic braking system is modelled in a simplified way. Depending on the position of the 3/3–way–solenoid valve, the pressure build–up in the brake cylinder is delayed by first–order systems (Figure 9).

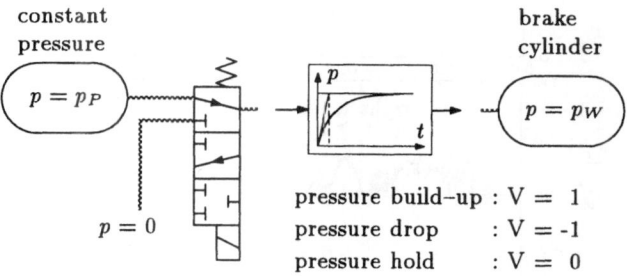

Figure9.: Simplified model of the brake–hydraulics

If the position of the solenoid valve is held constant during the control cycle a very large fluctuation of the pressure in the brake cylinders will arise. This is due to the very small time constants of the hydraulic system. An improved behaviour can be obtained by applying a pulse–width modulation (see Prochnio 1987). This method makes it possible to use intermediate values for the position V of a solenoid valve. For example $V = 0.6$ means that during the first 60 % of the controller–cycle time the pressure is built up, while during the remaining time of the cycle the pressure is held.

4   Simulation Results

This section demonstrates the operation of the fuzzy traction control system with simulation results for the model of a Mercedes-Benz S300. Especially on icy road conditions the operation of an ASR leads to a remarkable improvement of stability at high velocities. In this manoeuvre the vehicle accelerates on an icy road ($\mu = 0.2$) with a starting velocity of $3m/s$. Coevally, after $t = 2$ s the steering-angle changes to 5 °.

The consequence is that in this manoeuvre the vehicle reaches *low, medium* and *high* velocities. By this way it is possible to demonstrate the mode of operation of the single braking–torque control, the single engine–torque control but also of the combination of both at *medium* velocities.

94

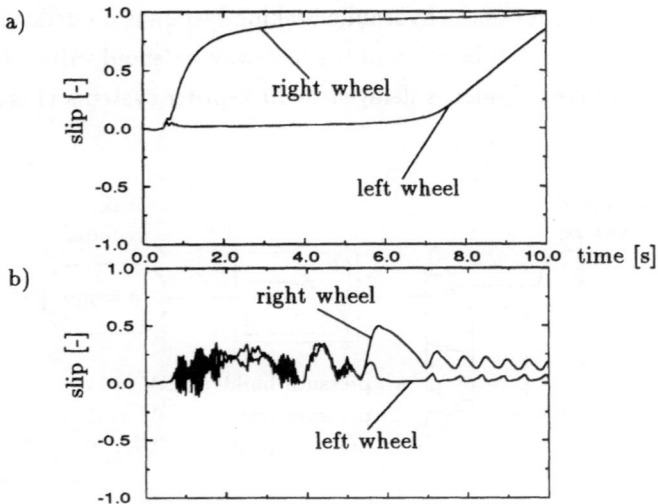

Figure10.: Simulation of a step steering manoeuvre on an icy road:
a) non-controlled,
b) controlled

Obviously, during the first 3s only the braking–torque control is active. Due to
the acceleration process, the vehicle reaches *medium* velocities between $t = 3s$
and $t = 6s$. Now the braking–torque control as well as the engine–torque control
are active. After $t = 4.0\ s$ the braking–torque control is inactive, because the
vehicle has reached a *high* velocity where only the position of the servo throttle
valve is controlled. The outputs of the controller can be taken from Figure 11.

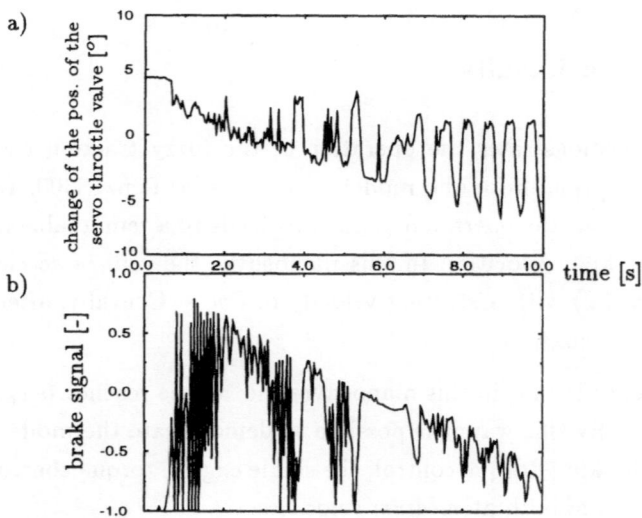

Figure11.: Combined control:a) engine–torque control,b) braking–torque control

The influence of the controller on the vehicle brakes decreases, while the influence of the ASR on the servo throttle valve increases. In this simulation the combined control strategy used in this controller leads to a remarkable improvement of traction and stability at the same time (Figure 12).

For further simulation results concerning the fuzzy traction control system the reader is referred to Schuster 1993.

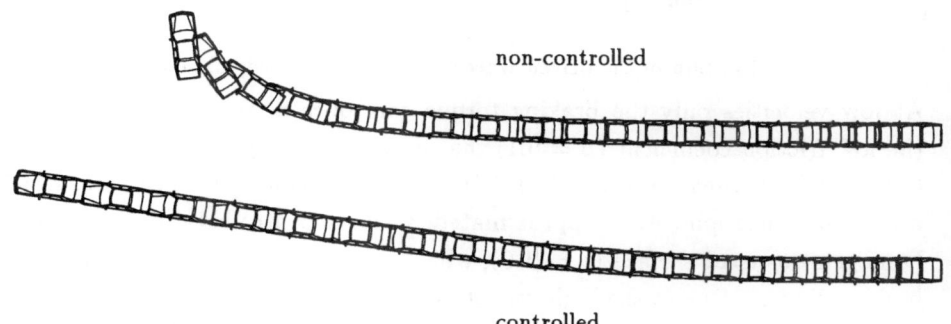

non-controlled

controlled

Figure12.: Trajectories of a controlled and a non-controlled vehicle

5   Stability and Robustness of the Controller

Due to the high complexity and nonlinearity of the system to be controlled it is impossible to investigate the stability of the system with common methods of control theory. Instead, the stability and robustness of the traction control system will be investigated by a systematic variation of the parameters of the vehicle.

As mentioned before, the two main requirements of an ASR are to maintain the stability of the car and to improve its traction especially on bad road surface conditions. To investigate the stability of the controller, two characteristic simulation manoeuvres were chosen.

In the first simulation an acceleration process on μ-split conditions ($\mu = 0.1 / 1.0$) starting with a velocity of 1.2 m/s will be considered. Figure 13 shows the slip of the driven wheels with respect to time for a controlled and a non-controlled driving manoeuvre. These results were measured for a vehicle with the original weight of 1840kg.

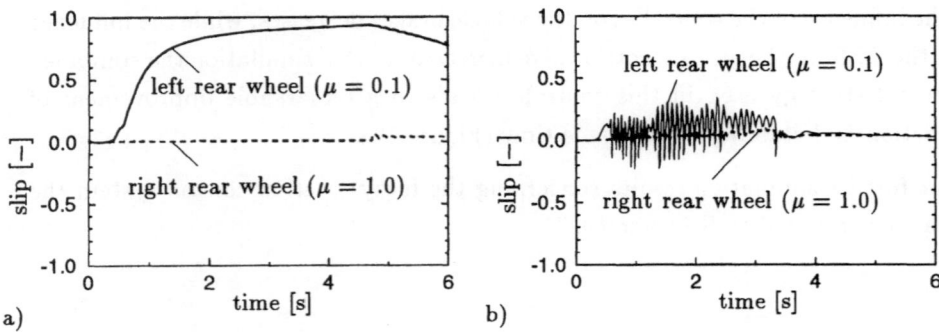

Figure13.: Slip of the driven wheels: a) non–controlled, b) controlled

At low velocities only the braking torque control is active. On the side with the low friction coefficient ($\mu = 0.1$) the slip is limited by the controller to 0.3 (30 %) while it runs up to nearly 1.0 (100 %) in the non–controlled case, which means full wheelspin. After approximately 3.5 s the controlled vehicle reaches a homogeneous road surface ($\mu = 1.0$), which causes a fast pressure drop in the brake cylinders. Due to the high velocity in both cases no significant yaw angles of the cars can be detected. The improvement of traction can be shown by means of the covered distance in a fixed time interval.

The improvement of the stability of the vehicle can be demonstrated with the help of a step-steering manoeuvre on an icy road. Here the sideslip angle β (Figure 14) between the longitudinal median plane of the vehicle χ and its horizontal velocity v_h serves as a measure for the stability and steerability of the vehicle (see Berkefeld 1991).

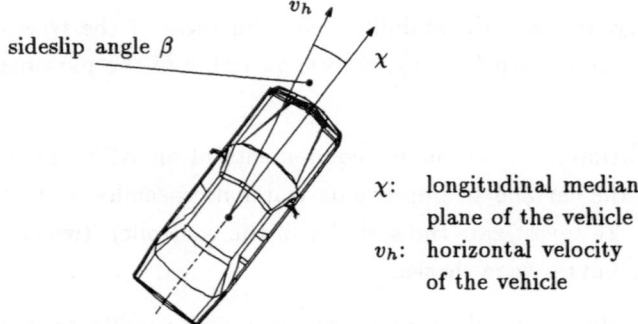

χ: longitudinal median plane of the vehicle
v_h: horizontal velocity of the vehicle

Figure14.: Definition of the sideslip angle

After $t = 2$ s, the steering-angle changes to $5°$. The friction coefficient changes at $t = 1$ s from 1.0 to 0.1 on both sides of the road and the vehicle accelerates with a starting velocity of $v = 10 m/s$. With the developed engine torque control a limitation of the logitudinal slip to less than 0.1 (10 %) is reached (see Figure 15a). Obviously, the braking torque control is characterized by a better sensitivity than the engine torque control. (see Figure 13). These results were also measured for a vehicle with the original weight of 1840kg.

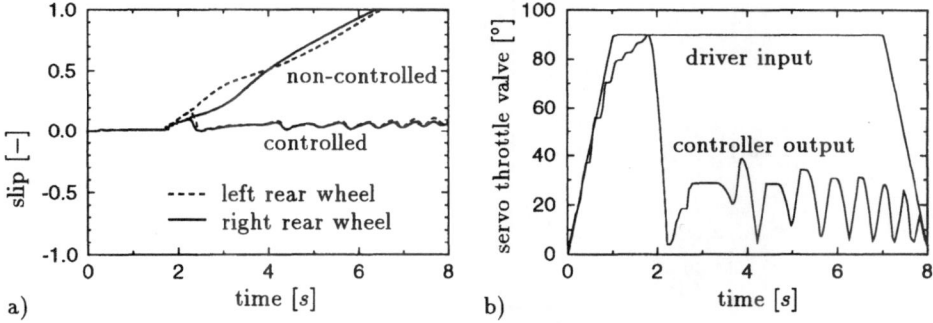

Figure15.: Simulation of a step steering manoeuvre on an icy road

In this investigation a large amount of parameters of the dynamical vehicle model and of the controller have been varied. It is well known that a fuzzy logic controller is robust towards an uncertainty in a large variety of cases. As an example the variation of the vehicle weight will be demonstrated in this paper.

In practice the vehicle weight is probably the mostly varied parameter. Figure 16a shows the sensitivity of the braking torque control towards a variation of the vehicle weight. Due to the increasing vehicle weight, the distance covered after t=6s decreases, but in the entire area of variation the controller remains stable. The robustness of the engine torque control can be taken from Figure 16b.

a.)

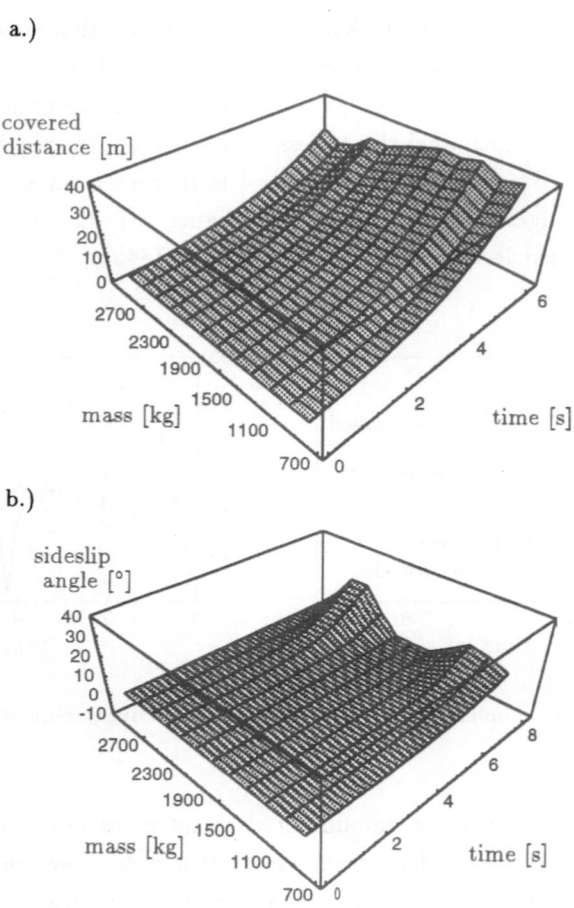

b.)

Figure16.: Variation of the vehicle weight

The second figure shows two extreme values of the sideslip angle at a vehicle weight of 1100kg and of 2300kg. The variation of any of the other parameters has shown that the controller is able to improve the stability until a physical limit is reached. Here this is not the case. It is quite surprising that a vehicle with a weight of almost 3000kg is easier to control than a vehicle with a weight of 2300kg. This interesting effect is currently under investigation.

The development process of the ASR has shown, that a first "raw" adjustment of the controller is easy to manage. The main difficulty is the second step, the optimization of the controller. As an example the optimized controller for the Mercedes Benz S300 was used unchanged with the vehicle model of a VW Golf. Figure 17 shows the trajectory of both vehicles for a step-steering manoeuvre at a high velocity.

a.)

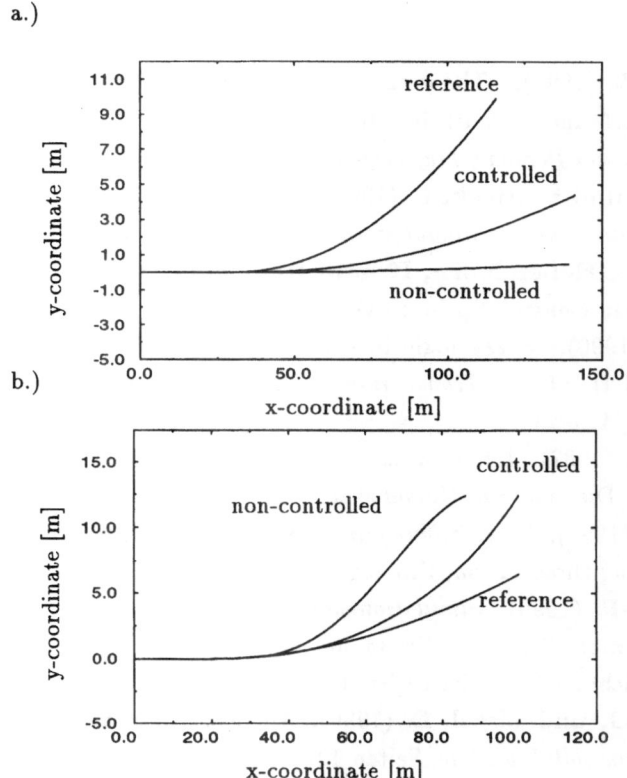

b.)

Figure17.: Trajectories of a front (a) and a rear driven vehicle (b)

The dynamical behaviour of the non-controlled vehicles in this manoeuvre is different due to the different concepts of power trains. The front wheel driven VW Golf tends to understeer while the rear wheel driven Mercedes Benz S300 oversteers. Nevertheless in both cases the ASR improves the steerability of the vehicles and demonstrates its robustness towards a change of parameters.

6   Conclusions

In this paper the design of a fuzzy traction control system and a first investigation on its stability and robustness have been presented. The development of the controller was managed with the help of spatial vehicle dynamics simulation. The controller has proved to be robust towards a variation of a large variety of parameters, in all simulations no instabilities were observed. It was even possible to connect a controller to a vehicle with a different power train concept. The global mode of operation of the fuzzy traction control system is always warranted, but in each different vehicle it has to be optimized and adapted.

References

Berkefeld, V. (1991). Theoretische Untersuchungen zur Vierradlenkung Stabilität und Manövrierbarkeit. In Wallentowitz, H. (Herausgeber), *Allradlenksysteme bei Personenkraftwagen*, Seiten 1–20.

Bertram, T. und Svaricek, F. (1992). Zur Fuzzy-Regelung eines aufrechtstehenden Pendels. *at – Automatisierungstechnik*, 40(8):308–310.

Driankov, D., Hellendoorn, H. und Reinfrank, M. (1993). *An Introduction to Fuzzy Control*. Springer-Verlag.

Lee, C. C. (1990). Fuzzy logic in control systems: Fuzzy logic controller — Part I and II. *IEEE Transactions on Systems, Man, and Cybernetics*, 20(2):404–418, 419–435.

Prochnio, E. (1987). *Ein Konzept zur pulsmodulierten Regelung hydraulischer Antriebe*. Dissertation, Universität–GH–Duisburg.

Schmitz, T. (1993). *Modellbildung und Simulation der Antriebsdynamik von Personenwagen*. Dissertation, Universität–GH–Duisburg.

Schnelle, K.-P. (1990). *Simulationsmodelle für die Fahrdynamik von Personenwagen unter Berücksichtigung der nichtlinearen Fahrwerkskinematik*. Fortschrittberichte VDI Reihe 12 Nr. 146. VDI-Verlag, Düsseldorf.

Schulze, B.-G. und Lissel, E. (1986). Antriebsschlupfregelung im IRVW 3 (ASR). *Automobil-Industrie*, Seiten 137–143.

Schuster, C. (1993). Auslegung einer Antriebsschlupfregelung (ASR) mit Hilfe der Fuzzy-Logik. Diplomarbeit, Fachgebiet Mechatronik, Universität–GH–Duisburg.

Zadeh, L. A. (1973). Outline of a new approach to the analysis of complex systems and decision processes. *IEEE Transactions on Systems, Man, and Cybernetics*, SMC-3(1):28–44.

Zimmermann, H.-J. (1993). *Fuzzy Technologien*. VDI-Verlag, Düsseldorf.

Knowledge-Based Rules for Control of the Sake (*Ginjoshu*) Making Process and Their Application in Fuzzy Control

H. HONDA[1], T. HANAI[1], Y. NISHIDA[2], I. FUKAYA[2] and T. KOBAYASHI[1]

[1]Department of Biotechnology, Faculty of Engineering, Nagoya University, Chikusa-ku, Nagoya 464-01, Japan

[2]Food Research Institute, Aichi Prefectural Government, 2-2-1 Shinpukuji-cho, Nishi-ku, Nagoya 451, Japan

ABSTRACT

For temperature control of excellent sake (*Ginjoshu*) making process, the construction of the knowledge-based rules was studied   From the interview with process operators (*Toji*), 4 knowledge based rules were listed up. Especially, it was found that Alcohol-Baumé plot and BMD curve were important as a reference for making *Ginjoshu* with good quality. Fermentation period was separated into 4 control regions. Fuzzy production rules and membership functions for temperature control in each region were constructed by using above references. Tuning of membership function was done using the simulator as proposed by us and the brushed-up rule was applied to 25L and 250L-*Ginjoshu* fermentation. Time course of temperature, acidity, Baumé, glucose, pyruvic acid, some alcohols and some esters were judged to be similar to those of traditional fermentation process by *Toji*.

1. INTRODUCTION

Ginjoshu making process is one of Japanese traditional fermentation process. *Toji* are experts for this process and they have many precise know-hows and rules based on their experience. Recently, however, the aging of their population and the decrease of the young successor have caused a crisis for sake brewing companies.

To solve this problem, the development of labor-saving process has been required to engineers from both software [1] and hardware side. Especially for fermentation (*moromi*) process, it is needed to construct the software based on many know-hows and rules, from which the strategy for process control could be decided. In this paper, a fuzzy resonance was applied to construct the strategy for temperature control, which is the most important variable in *Ginjoshu moromi* process.

2. MATERIALS AND METHODS

58 data sets of time course of *Ginjoshu* brewed between from the year 1989 to 1991 in Aichi prefecture in Japan was used for data analysis. The data sets consist of temperature, Baumé, and alcohol concentration which were collected everyday. Baumé is a technical term used in sake making process, which corresponds to a specific gravity of fermentation broth, *moromi*. Software of fuzzy control was programed by N88-Basic language using a personal computer (PC-9801 RA2; NEC Co.Ltd.,Tokyo) .

For the experimental fermentation, *Saccharomyces cerevisiae* FIA-2 was used as a microorganism. As a raw material, 10 kg of steamed rice and 15L of water were used for 25L fermentation.

3. RESULTS AND DISCUSSION

3.1 ANALYSIS OF FERMENTATION DATA

Figure 1 shows time courses of temperature, Baumé and alcohol concentration of 58 *moromi* fermentations. Temperature distribution was narrow in the early period and became wider in the later period. This means that the degree of temperature decreasing at later period was changed widely in each of batch processes depending on the initial condition or the fermentation state such as the rate of alcohol fermentation and saccharization. Average temperature of the first day was 6.1°C. The highest temperature was attained after the 10th day and its average was about 9.8°C. The day corresponded to the day that the foam on the broth disappeared (the *bouzu* day), after that temperature had been lowered slowly until about 5 °C.

Figure 2 shows time course of difference in temperature (ΔT) between the day and the previous

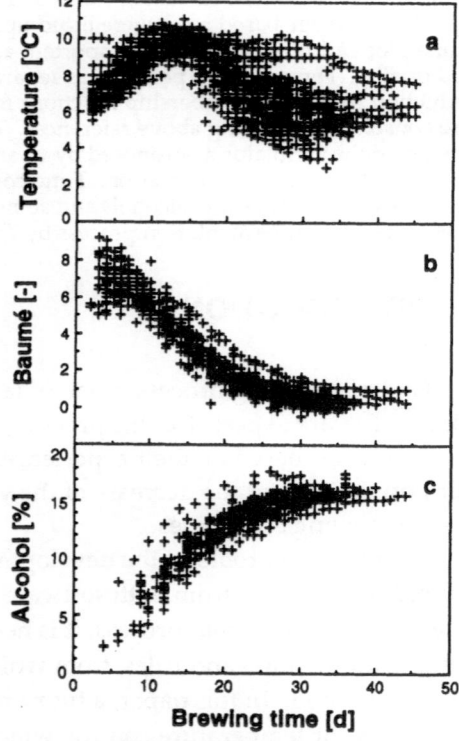

Fig. 1 Profiles of fermentation parameters; temperature (a), Baumé (b) and alcohol concentrarion (c), in 58 *ginjyo moromi* fermentations.

day. From Fig.2, it was found that temperature was controlled as follows; only raised from the first day to the *bouzu* day, only lowered from about the *bouzu* day to the 22nd day and both raised and lowered from about the 22nd day to the last day. It was also found that the *bouzu* day was the day that alcohol concentration was over 10.5% and Baumé was under 3.5 and the 22nd day was almost equal to the day that Baumé become under 2.0.

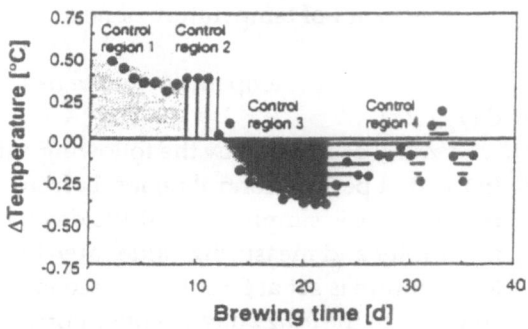

Fig.2 Time course of average values of ΔTemperature

3.2 KNOW-HOWS FROM *TOJI* [2] AND KNOWLEDGE-BASED RULES STRATEGY FOR TEMPERATURE CONTROL [3]

From the interview with *Toji* , the following knowledge-based rules for making *Ginjoshu* with good quality were listed. 1) *Moromi* fermentation should be finished at about 30day. 2) Keeping low temperature at end of *moromi* is important. 3) Temperature should be controlled based on straight line of Alcohol-Baumé plot after 9th day, and 4) based on straight line of BMD (Baumé Multiplied Day)-Day plot in the later period. In order to confirm the third and forth rules, we analyzed 58 data sets. As shown in Figure 3, it was found that the average data of A-B and BMD-D plot could be put on a straight line. Based on these rules and results from Fig. 2, we constructed the knowledge-based rules for temperature control.

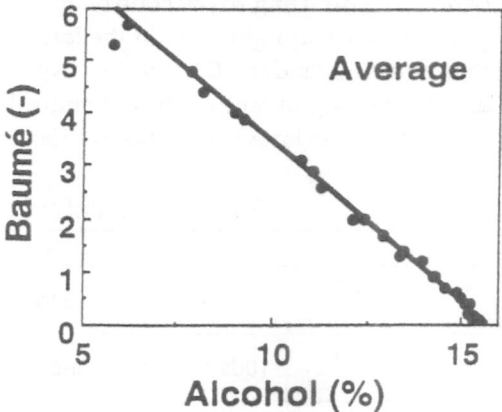

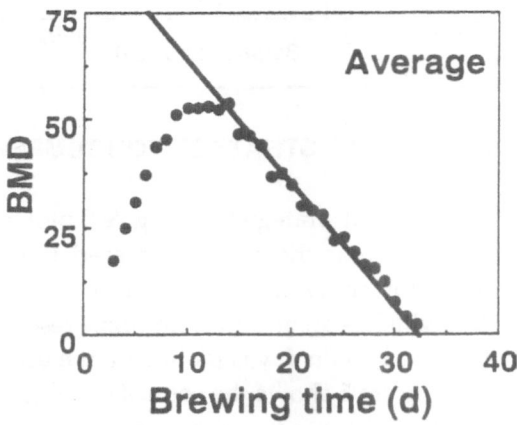

Fig. 3 Alcohol - Baumé plot and BMD - D plot of 58 data sets of actual *moromi* fermentation

Set point of temperature of next day was determined by the following equation.

(Temperature of the n+1 st day) = (Temperature of the n th day) + $\triangle T$

$\triangle T$ was decided by the following rules based on the knowledge of *Toji*. In the first period (control region 1 : from the first day to the 9th day),. temperature is simply raised 0.4°C day by day, since fermentation broth has no fluidity and measuring values aren't reliable as process data. Initial temperature is set at 6°C. In the second period (control region 2 : from the 10th day to the *bouzu* day), temperature is only raised in order that A-B plot adjusts to that of average values. $\triangle T$ was decided from fuzzy rules. In the third period (control region 3 : from the *bouzu* day to the day that Baumé is over 2.0), also using fuzzy control, temperature is only lowered so that A-B plot follows a straight line. In the last period (control region 4 : from the day that Baumé is under 2.0 to the last day), using fuzzy control, both increasing and decreasing of temperature is controlled so that BMD value decreased linearly. These knowledge-based rules are summarized in Table 1.

Table 1 Control strategy and reference.

	period	operation of temperature	reference
Control region 1	1-9day	simply increasing (0.4°C/d)	—
Control region 2	10day-Bouzu	increasing by Fuzzy control	straight line on A-B plot
Control region 3	Bouzu-Baumé over 2.0	decreasing by Fuzzy control	straight line on A-B plot
Control region 4	Baumé under 2.0	increasing or decreasing by Fuzzy control	straight line on BMD curve

3.3 ACTUAL STRATEGY FOR TEMPERATURE CONTROL

Control strategy by using A-B plot was described in Fig. 4. If temperature in the tank is increased, the decrease of Baumé is promoted. As a result, lower value of Baumé will be obtained at the next day. In the other words, $\triangle T$, which is an output from fuzzy resonance, should become positive, if the data of a certain day is plotted on the upper side of reference.

Figure 5 shows the control strategy by using BMD-D plot. In the region 4, BMD reference was drawn according to the data in the region 3. If the cross point of the reference with X-axis is out of the range between 29 to 33 day, the reference is redrawn as it is into this range. $\triangle T$ is basically controlled as the same as the strategy in Fig. 4. Out put, $\triangle T$, is positive, if the data of a certain day is plotted on the upper side of reference. Membership function and production rules of the region 2 and 3 and region 4 were described in Figs. 6 and 7, respectively. Production rules were constructed by using the ratio of length and △degree in Fig. 6 and temperature and △BMD in Fig. 7.

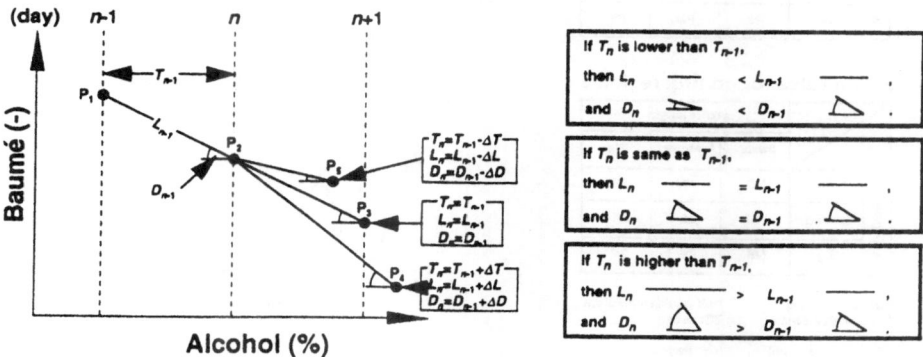

Fig. 4   Control strategy for *Ginjo moromi* fermentation on A - B plot

In the region 2, all of production rules are zero or positive, since temperature in this region should be increased; if the ratio of length is big and △degree is big, then the production rule is positive big. On the other hand, in the region 3, all of production rules are zero or negative, since temperature in this region should be decreased; if the ratio of length is small and △degree is small, then the production rule is negative big. In the region 4, there are negative, zero and positive of the production rule. If the △BMD is big and temperature is small, then the production rule is positive small in order to increase temperature slightly.

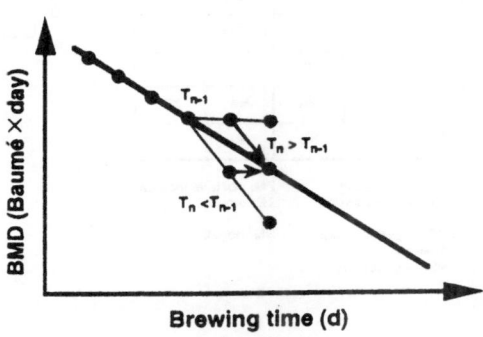

Fig.5   Control strategy for temperature control based on straight line of BMD curve.

Production rules for control region 2

	ratio of length		
	Small	Medium	Big
Small	ZE	ZE	PS
Medium	ZE	ZE	PM
Big	PS	PM	PB

(Δ Degree)

Membership functions for inputs

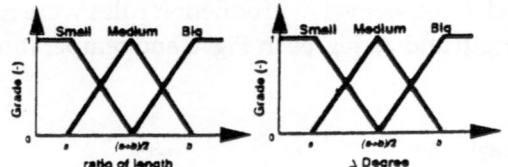

Production rules for control region 3

	ratio of length		
	Small	Medium	Big
Small	NB	NM	NS
Medium	NM	ZE	ZE
Big	NS	ZE	ZE

(Δ Degree)

Membership functions for outputs

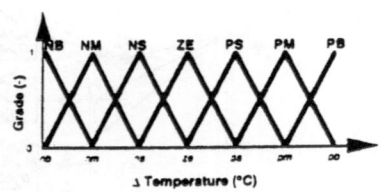

PB: positive big   PM: positive medium
PS: positive small   ZE: zero

NS: negative small   NM: negative
medium
NB: negative big.

Fig. 6   Production rules and membership functions for control region 2 and 3

Membership functions for inputs

Production rules for control region 4

	Δ BMD		
	Small	Medium	Big
Small	NS	ZE	PS
Medium	NM	NS	ZE
Big	NB	NM	NS

(Temperature)

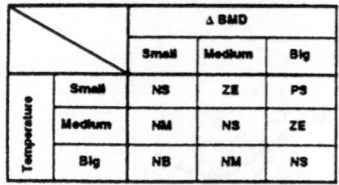

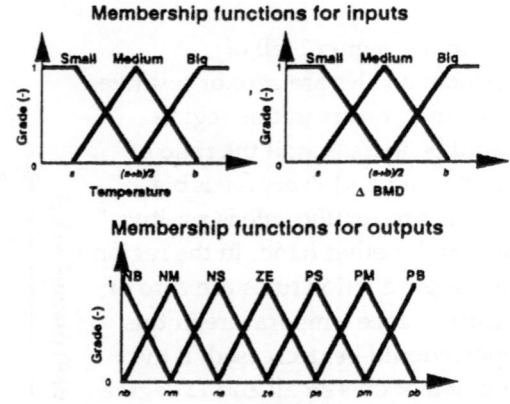

Membership functions for outputs

PB: positive big   PM: positive medium
PS: positive small   ZE: zero

NS: negative small   NM: negative
medium
NB: negative big

Fig. 7   Production rules and membership functions for control region 4

Some parameters were tuned by the calculation using the simulator. This simulator was constructed by us based on the relationship between the integrated temperature and the alcohol concentrations or Baumé (formulation was not shown). Some parameters of membership function determined by the calculation are listed in Table 2.

Table 2   Parameters of membership functions

		a	b
Control region 2	ratio of length	0.5	1.5
	△degree	-15	15
Control region 3	ratio of length	0.5	1.5
	△degree	-15	15
Control region 4	△BMD	-2.0	2.0
	temperature	5.0	8.0

		nb	nm	ns	ze	ps	pm	pb
△Temperature	Region 2	-	-	-	0.0	0.1	0.2	0.3
	Region 3	-0.3	-0.2	-0.1	0.0	-	-	-
	Region 4	-0.7	-0.5	-0.2	0.0	0.2	-	-

3.4 TUNING OF THE FUZZY RULES

The suitability of fuzzy rules was confirmed from the simulation under various initial conditions, in which various 9 days-Baumé values (maximum Baumé value) were used. Results are described in Table 3. The calculated results were coincided well with the actual experimental data.

Table 3   Comparison of the calculated values with the average values of actual fermentation

		Brewing time (day)					
		10	15	20	25	31	32
Temperature (C)	Calculation	9.3	9.5	7.6	5.9	5.5	5.4
	Measurement	9.3	9.3	7.5	6.1	5.2	5.2
	Difference	0.0	0.2	0.1	- 0.2	0.2	0.2
Baumé (-)	Calculation	5.3	3.2	1.6	0.8	0.2	0.0
	Measurement	5.3	3.1	1.7	0.9	0.2	0.1
	Difference	0.0	0.1	- 0.1	- 0.1	0.0	- 0.1
Alcohol (%)	Calculation	6.5	10.4	12.8	14.3	15.5	15.6
	Measurement	5.8	10.7	12.9	14.3	15.3	15.3
	Difference	0.7	- 0.3	- 0.1	0.0	0.2	0.3

3.5 25L-*GINJOSHU* EXPERIMENTAL FERMENTATION

Brushed-up rule was applied to 25L-*Ginjoshu* fermentation. Figures 8, 9 and 10 show the time course of temperature, A-B plot and BMD curve of this fermentation, respectively. Figure 8 shows that temperature was increased and decreased according to each control rule. As shown in Fig.9, A-B plot was almost straight in control region 2 and 3. BMD curve was also straight in the control region 4 (Fig.10). This means that temperature as an operational parameter was set fairly well and both alcohol and Baumé could be enough controlled.

3.6 250L-*GINJOSHU* FERMENTATION

Using the fuzzy rule, actual scale fermentation of 250L containing 100 kg-rice was performed. The result was compared with control fermentation operated in manual by expert. Figure 11 shows the time course of temperature, Baumé and alcohol concentration. In the region 4, temperature in fuzzy control was observed to be slightly higher than that in manual operation. This may be due to the slight difference of fermentation

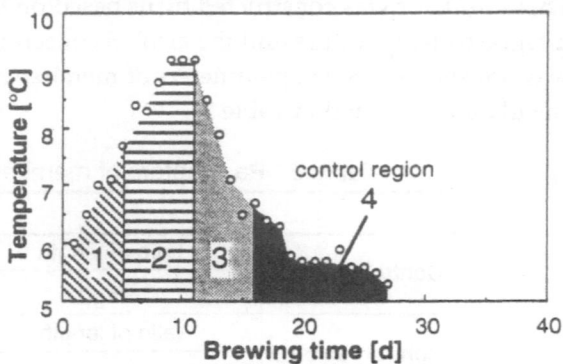

Fig.8   Time course of temperature in 25L-experimental fermentation

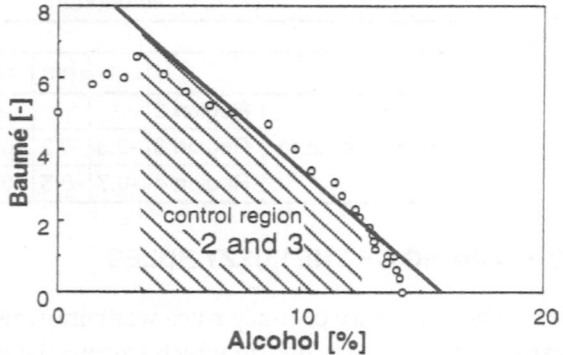

Fig.9   A-B plot in 25L-experimental fermentation

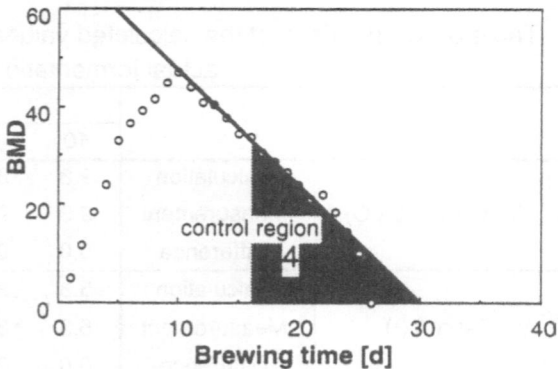

Fig.10   BMD curve in 25L-expermental fermentation

condition such as mixing. However, it was found that in all regions the profiles of Baumé and alcohol concentration were almost similar to those of control culture. This means that using these rules constructed by us, *moromi* temperature was almost controlled similar to the control of *Toji* .

Figure 12 shows the A-B and BMD-D plot in these fermentations. Data points from two fermentations were found to be put on the straight line. From Fig. 12, it was confirmed that the temperature could be operated based on the fuzzy rule and it results in the same time course as that in manual operation by expert, *Toji*.

Table 4 shows some analytical data of *Ginjoshu* from 250L-fermentation. All analytical data were found to be almost similar to those in manual operation. Since in general those data are strongly related to the quality of *Ginjoshu*, the *Ginjoshu* made by the fuzzy control was concluded to have a good

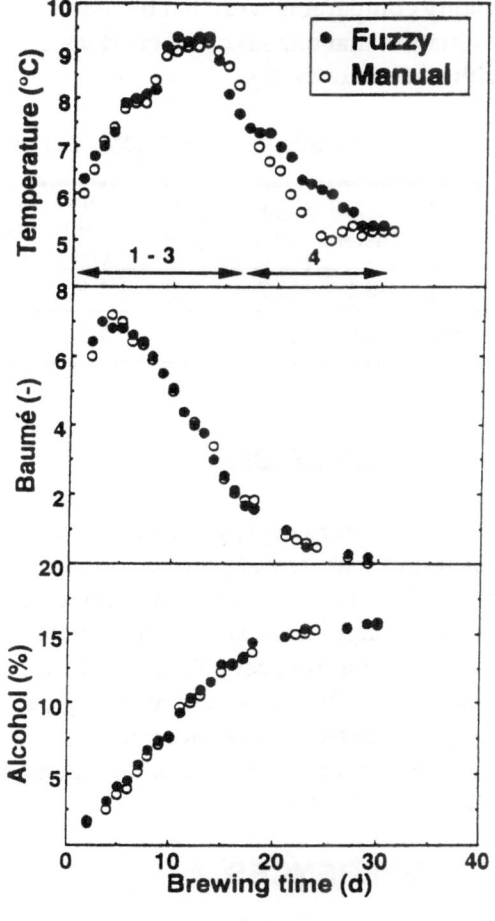

Fig. 11 Time course of 250L-fermentation by fuzzy control and manual control.

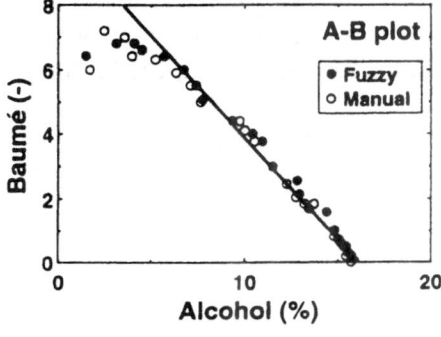

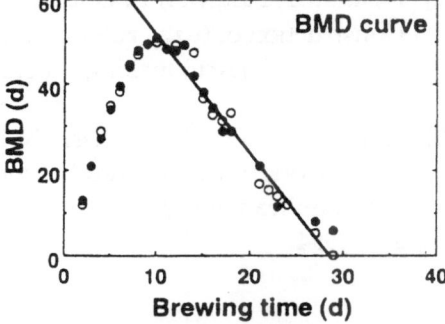

Fig. 12   A - B and BMD - D plot of 250L - *moromi* fermentation

quality comparable with that by manual control. It was confirmed by sensory evaluation that our sake had good taste and flavor and it was comparable with one made by *Toji*.

Table 4 Comparison of analytical data of products, *Ginjoshu*.

	Sake meter	Alcohol (%)	Acidity (ml)	Amino acidity (%)	Total sugar (%)	iAmOH (ppm)	iAmOAc (ppm	E/A x 100	Et. Cap (ppm)
Fuzzy	-0.2	15.7	1.6	1.3	50.3	112.8	2.23	1.97	2.3
Manual	0.1	15.8	1.6	1.3	47.1	127.2	2.51	1.97	2.3

4. CONCLUSIONS

From interview with *Toji* and data analysis of 58 *moromi* fermentation, it was found that *moromi* process was separated to 4 control regions, A-B plot and BMD curve were important as indices for temperature control based on the knowledge-based rules. The fuzzy rules were constructed in order to decide temperature set point everyday. 25L and 250L *moromis* could be almost controlled under the fuzzy rule, and the *Ginjoshu* with good taste and flavor was brewed. Therefore, it was concluded that our strategies for temperature control of *moromi* fermentation process could become a useful software tool replacing *Toji*.

5. REFERENCES

[1] Oishi K, Tominaga M, Kawato A, Abe Y, Imayasu S, Nanba A, Application of fuzzy control theory to the sake brewing process. J. Ferment. Bioeng. vol.72, 115-121, 1991
[2] Nishida Y, Fukaya I, Takahashi N, Hanai T, Honda H, Kobayashi T, Construction of fuzzy rules based on statistically analyzed for control of the sake (*Ginjoshu*) making process. Seibutsu Kogaku Kaishi, vol.72, 267-274, 1994
[3] Hanai T, Honda H, Takahashi N, Nishida Y, Fukaya I, Kobayashi T, Framework rules for control of the sake (*Ginjoshu*) making process and their application in fuzzy control. Seibutsu Kogaku Kaishi, vol.72, 275-281, 1994

A Framework for Studying the Effects of Dynamic Crossover, Mutation, and Population Sizing in Genetic Algorithms

Michael A. LEE

University of California
Computer Science Division
Soda Hall
Berkeley, CA 94720-1776 USA
leem@cs.berkeley.edu

Hideyuki TAKAGI*

Kyushu Institute of Design
Dept. of Acoustic Design
4-9-1 Shiobaru, Minami-ku
Fukuoka, Fukuoka 815 JAPAN
takagi@kyushu-id.ac.jp

Abstract - We introduce a framework for controlling genetic algorithms and use it to study the effect of dynamically modulating genetic algorithm control parameters on search behavior. Our framework includes techniques that can automatically design control strategies for genetic algorithms according to a given search performance metric. Many of the automatically designed strategies exhibit an exponentially decreasing mutation rate behavior. We present experimental results indicating that exponentially decreasing the mutation rate over time contributes more towards an increase on online and offline search performance than populations size or crossover rate modulation.

1 Introduction

The relationship between a genetic algorithm's control parameters and its behavior is complex and has been the topic of much research. For example, the behavior of a genetic algorithm can range from that of random search to hill climbing depending on the genetic algorithm parameters settings. Designing a genetic algorithm that meets resource constraints of a given application may require a substantial amount of genetic algorithm control parameters tuning. In addition, because the genetic algorithm search is a dynamic process, maintaining a given search behavior may require dynamically modulating the control parameters. In this study, we present a framework for controlling genetic algorithm parameters and techniques to automatically design high performance strategies.

The remainder of the paper is divided into 7 major sections. The next section discusses related work and how our work builds on the results presented by others. Section 3 describes our framework for representing search control strategies: the Dynamic Parametric Genetic Algorithm (DPGA) -- a genetic algorithm controlled by a fuzzy knowledge-base. Section 4 presents our technique for automatically designing and tuning genetic algorithm control strategies. Sections 5 and 6 present experiments and results aimed at studying the interaction between mutation rates, crossover rates, and popula-

---

* Portions of this research were carried out while the author was a Visiting Industrial Fellow at UC Berkeley on leave from Central Research Laboratories, Matsushita Electric Industrial Co., Ltd.

tion sizing. In the final section, we explain how to extend our technique to other learning methods. We also outline areas that warrant further investigation.

2 Background and Related Work

Genetic algorithms were introduced and developed by John Holland and his colleagues as a system aimed at explaining and modelling adaptation of natural systems [11]. A basic genetic algorithm involves selection, mixing, and mutation components. Like biological systems, selection is driven by an organism's ability to survive in its environment. If the organism lives long enough to reproduce with other organisms, it will pass its genetic information onto its offspring. Although mature organisms are capable of reproducing, the most fit organisms usually have more opportunities to produce offspring. Mixing in genetic algorithms is usually implemented by crossover operators that combine genetic information from two or more parents. Mutation is a mechanism for reintroducing information that may not be contained in population. The population serves as a organism with distributed knowledge: knowledge is distributed throughout the genes of the entire population.

Recently, genetic algorithms have been removed from their original context and studied in a function optimization context. Much effort has gone into refining the original simple genetic algorithm to speed convergence and improve robustness. Two general approaches toward improving genetic algorithm performance have been 1) development of adaptive mechanisms within the genetic algorithm and 2) optimization of static parameters such as mutation rate or population size. Among the ideas that have been introduced are fitness scaling, generation gap, and rank selection.

Fitness scaling was introduced to adapt the selective pressure according to the worst member of a population. The fitness used for selection is computed relative to the fitness measures the rest of the population. As the entire population moves towards higher fitness values, the pressure to select the better solutions is preserved -- avoiding the evolution of a population of mediocre individuals. Generation gap is another mechanism to prevent premature convergence by allowing structures from the current generation to automatically advance to the next generation.

Baker investigated a selection scheme based solely on rank [3]. Members are sorted according to their fitness values and then a member's rank is used to compute the expected number of offspring it will contribute to the next generation. In this work, Baker compared his selection scheme to others. The results of this work did not show that this method was better than the others; however it did show immunity to the selection problems previously mentioned. For a more complete comparison of selection methods see [7].

Fogarty has studied the effects of a adaptive mutation operators [5]. In this work, several mutation operators were compared; constant mutation rate, mutation rate varied over generation, mutation rate varied over integer representation, and mutation rate varied over generation and integer representation. The results of his experiments

113

showed that dynamically changing the mutation rate always showed an improvement over the constant case.

Goldberg et al. investigated the effect of population size on GA behavior [8]. A population sizing equation was derived that considers variance of building-block fitness, noise of the genetic operators, and noise in the objective function. The authors determined that with small populations, the GA performance is determined by chance, by mutation, or by another mechanism that serially injects diversity. Among several extensions to this work that were outlined, an online population sizing technique was proposed. The online sizing of the population could be based on information about the problem size, population variance, minimum signal, and order of deception.

A number of researchers have proposed encoding meta-level knowledge directly into the genetic code. The meta-level parameters are subjected to genetic operations as are the application parameters. Bagley proposed including crossover and mutation probabilities in the genetic code [2]. Schaffer and Morishima investigate an adaptive crossover scheme in which crossover points are contained in an extended bit representation in [18]. An additional binary code, equal in length to the original genetic code, is appended to the original code. This extension contains marks or punctuations where crossover points are valid.

Perhaps the first to study genetic algorithms and compare them with conventional gradient techniques is De Jong [4]. In this study, De Jong proposed a set of five functions to represent a wide cross section of function families [4,6]. He designed two measures to quantify a search technique's performance: *online* performance to measure ongoing performance and *offline* performance to measure convergence. Online performance is the running average of all evaluations performed up to a given time:

$$x_{online}(s, e, T) = \frac{1}{T}\sum_{t=1}^{T} f_e(t) \tag{1}$$

where s is the search strategy, e is the environment, $f_e(t)$ is the objective function value at time t,. This measure may be appropriate in situations where the cost of evaluating a structure, or population member, is related in a monotonically increasing way to its fitness value (i.e., evaluating a poor solution is more expensive than evaluating a good one). Offline performance is the running average of the best performance value up to a given time:

$$x_{offline}(s, e, T) = \frac{1}{T}\sum_{t=1}^{T} f_e^*(t) \tag{2}$$

where $f_e^*(t)$ is the best function value obtained up to time t and T is the current number of evaluations. This measure can be used when there is no additional cost for evaluating poor structures.

De Jong then explored empirically the relationship between genetic algorithm performances, with respect to these measures, and genetic algorithm parameter settings. From these experiments, the following parameter settings for genetic algorithms were given as settings that give good online and offline performance on the test suite and were subsequently used as common settings: population size=50~100, crossover rate=0.6, mutation rate=0.001.

Grefenstette extended this work by using a meta-level genetic algorithm to find parameter settings that yield robust and high performance genetic algorithms [9]. The task of the meta-level genetic algorithm was to determine population size, crossover rate, mutation rate, window size, generation gap, and selection strategy settings that maximized either the online or offline performance over a range of applications. To measure the quality of a genetic algorithm, Grefenstette formulated a fitness function based on De Jong's online and offline performance measures. This fitness function normalized the genetic algorithm's performance measure on a given application environment with respect to the same performance measure obtained by random search on the same application environment:

$$fitness(s) = \sum_{i=1}^{N} \frac{x_{perf}(s, f_i, T)}{x_{perf}(rand, f_i, T)} \tag{3}$$

where x_{perf} represents the search performance measure, s represents a particular search strategy search strategy, f_i represents the application environments, and T represents the number of evaluations a search was allowed to execute. $rand$ represents a random search strategy. In Grefenstette's experiments, the five De Jong functions were used as the application environments.

His results gave parameter settings for best online performance which resulted to be population size=20~30, crossover rate=0.75~0.95, mutation rate=0.0005~0.01 for good online performance and population size=80, crossover rate=0.45, mutation rate=0.01 for good offline performance. In this work, Grefenstette also optimized parameter settings for window size, generation gap, and selection strategy and were given as (window size=7, generation gap=1.0, elite selection) for online and (window size=1, generation gap=0.9, pure selection) for offline respectively. (We will refer to these genetic algorithms as optimized static genetic algorithms.) These are static settings; the settings are entered at the beginning of the run and then are left unchanged throughout the run.

In our work, we extend Grefenstette's research by investigating methods that develop strategies for dynamically changing the genetic algorithm control parameter settings. In the next section, we describe the Dynamic Parametric Genetic Algorithm, our framework for representing genetic algorithm control strategies.

3 The Dynamic Parametric Genetic Algorithm

In this section, we present the Dynamic Parametric Genetic Algorithm (DPGA): a framework for representing genetic algorithm control strategies. The DPGA represents a class of genetic algorithms that uses a fuzzy knowledge-base to control its parameters dynamically (see Figure 1). In this figure, a fuzzy inference engine controls the

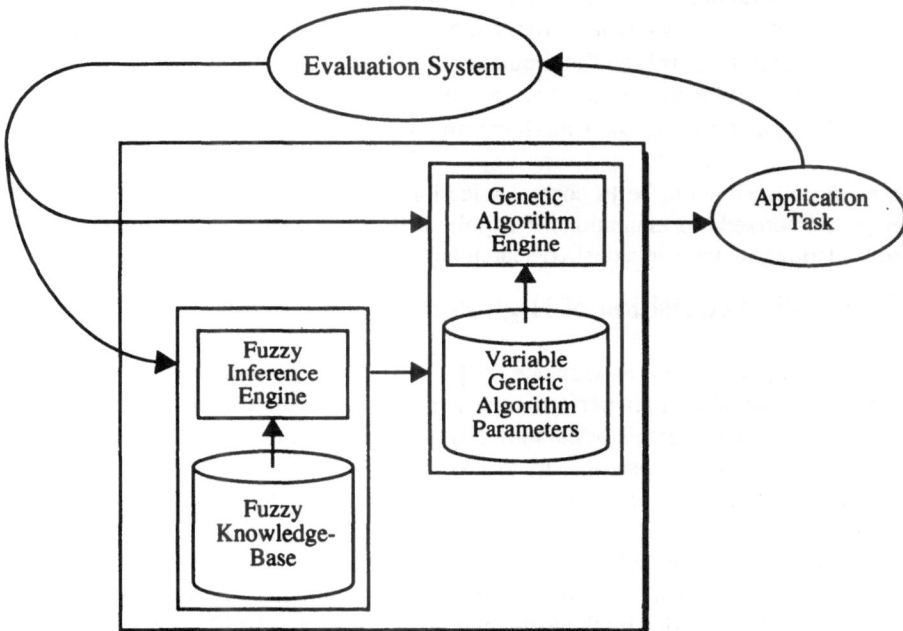

Fig. 1. The Dynamic Parametric GA framework: a genetic algorithm with dynamic parameters controlled by a fuzzy knowledge-based system. The fuzzy knowledge based system monitors performance measures from the evaluation system to control genetic algorithm parameters such as population size, mutation rate, or crossover rate.

parameters of a genetic algorithm, such as population sizing and mutation rate. The inputs to the fuzzy system are measures of the genetic algorithm search performance, which we assume can be derived from the quality of the solutions the genetic algorithm generates. The evaluation system is responsible for assigning fitness scores to solutions generated by the genetic algorithm.

Our decision to represent strategies using a fuzzy knowledge-base is predicated on three main points; fuzzy systems:

1. are rich enough to represent a broad range of strategies,
2. facilitates the capture and representation of expert knowledge through the use of linguistic rules,
3. can be automatically tuned and design using any number of numerous methods found in the literature.

Using the expressive richness of fuzzy systems gives us the ability to represent exist-ing control strategies within a single framework. Because of this generality, complex combinations of heuristic strategies can be implemented and compared. With auto-matic tuning and design techniques available for fuzzy systems, expert knowledge contained in the system can be both refined and augmented. For example, dynamic heuristic strategies of the following form may be represented within our framework:

- **if** generations is low **then** mutation rate is high
- **if** generations is medium **then** mutation rate is medium
- **if** generations is high **then** mutation rate is low
- **if** diversity is low **then** increase mutation
- **if** population average is not increasing **then** increase population size.

Other possibilities could be to control selection of operators and their parameters or even parameterized representations. The point is that the framework we propose is able to self-adapting to achieve a desired search behavior.

4 Automatic Acquisition of High Performance Control Strategies

In this section, we present an instance of the class of DPGA and demonstrate how to automatically acquire high performance genetic algorithm control strategies. The example DPGA has dynamic population sizing, crossover rate, and mutation rate. The fuzzy knowledge-based system used to control the DPGA outputs control actions in the form of percent change from current value: change in population size, crossover rate change, and mutation rate change. The ranges of the outputs are set such that the control parameters could not decrease by more than half their current value and could not increase be more than twice their current value. Population size, mutation rate, and crossover rate are bounded by the following ranges respectively: [2,160], [0.0001,1.0], and [0.0001,1.0]. Because the outputs represent changes in values as opposed to abso-lute values, the initial values of these parameters must be specified. The inputs to fuzzy knowledge-base are population diversity measures in the form of best to average ratio, average to worst ratio, and change in best solutions since the last generation (see Figure 2):

$$x_0 = \frac{\text{average fitness}}{\text{best fitness}} \tag{4}$$

$$x_1 = \frac{\text{worst fitness}}{\text{average fitness}} \tag{5}$$

$$x_2 = \Delta\text{best fitness since last control action} \tag{6}$$

Our technique for automatically acquiring control strategies is based on a fuzzy system design technique proposed by Lee and Takagi [15,16]. This technique uses genetic

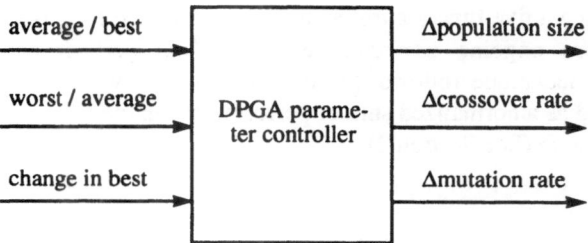

Fig. 2. Input and outputs to fuzzy knowledge-based system used in our Dynamic Parametric GA (DPGA) experiments. Inputs measure characteristics of the performance of the genetic algorithm on the environment. Outputs control the population size, crossover rate, and mutation rate.

algorithms to design membership functions, consequent parameters and number of rules of a fuzzy systems. To use this technique, we must design both a genetic representation of the DPGA and a fitness measure. These two points will be covered in the next two subsections.

4.1 DPGA System Parameterization and Genetic Coding

Our example DPGA requires two types of specifications: initial condition values for the three control parameters and a fuzzy system. Coding the initial value parameters is straight forward. For the fuzzy knowledge-base each of the three input and three output intervals in our example system can be covered a maximum of three fuzzy sets. The sets are constrained to overlap in such a way that exactly two of the membership functions have non-zero values and three sum of their memberships is 1. The maximum number of rules is determined by the number of output variables multiplied by the maximum number of input set combinations (assuming exactly one set per input dimension is present in the antecedent part of each rule). Each rule has exactly one consequent; one output set can be associated with a given input set combination in the antecedent.

The genetic coding of our DPGA follows naturally from the parameterization. We use a binary coding that includes alleles for initial conditions and alleles for the fuzzy knowledge-based system. The fuzzy knowledge-based system has alleles that determine membership function shape and rule structure [17]. The rule structure is designed such that an allele determines whether which output set, if any, is used as the consequent of a particular inputs set combination.

The fuzzy system uses triangular membership functions, the *min* intersection operator and correlation-product inference procedure. Defuzzification of the outputs is performed using the fuzzy centroid method [13].

4.2 Genetic Algorithm Search Performance Measures

The control strategies learned in our experiments were optimized according to performance measures and a five function test suite proposed by De Jong [4]: *online* performance to measure ongoing performance and *offline* performance to measure convergence. Our technique follows that of Grefenstette's where performance of a DPGA is measured as a normalized sum of performance on the De Jong functions relative to random search (See Section 2).

5 Results

We design separate Dynamic Parametric GAs for optimizing online and offline performance measures. In this section, we will look at the dynamic behavior of the DPGAs and compare the results with a simple static genetic algorithm proposed by De Jong (SSGA)[4], the optimized static online and offline genetic algorithms proposed by Grefenstette (OSGA)[9], and random search (see Table 1 for GA parameter settings).

Table 1: Genetic algorithm search parameter settings for simple static GA (SSGA), optimized static GA for online performance (OSGA online), and optimized static GA for offline performance (OSGA offline).

parameter	SSGA	OSGA online	OSGA offline
population size	50	30	80
crossover rate	0.6	0.95	0.45
mutation rate	0.001	0.01	0.01
generation gap	1.0	1.0	0.9
window size	7	1	1
selection strategy	Elite	Elite	Pure

The genetic algorithm used to design Dynamic Parametric GAs itself had fixed parameters of population size=10, crossover rate=0.8, mutation rate=0.0333. It used an elitist selection strategy and window sizes and generation gaps were fixed at 1 and 1.0 respectively. This genetic algorithm was allowed to evaluate 1000 Dynamic Parametric GAs.

We would like to caution the reader that the specific instances of DPGAs presented in the following section represent only one instance of the class of DPGAs. The DPGAs presented exhibit high performance, however it is very likely that other fuzzy rule sets may exhibit equal or better performance. The point of this exposition is to demonstrate the technique for automatically obtaining genetic algorithm control strategies.

5.1 Dynamic Parametric GA for Online Performance

The initial population of fuzzy systems and initial conditions used for determining the good online performance was seeded with an individual identical with the best online performing genetic algorithm determined by Grefenstette. In this individual, the population size was set to 30, and the crossover mutation rates were set to 0.95 and 0.01 respectively. For each fuzzy system produced for the Dynamic Parametric GA, the generation gap, window size, and selection strategy were fixed at 1.0, 1, and Elitist.

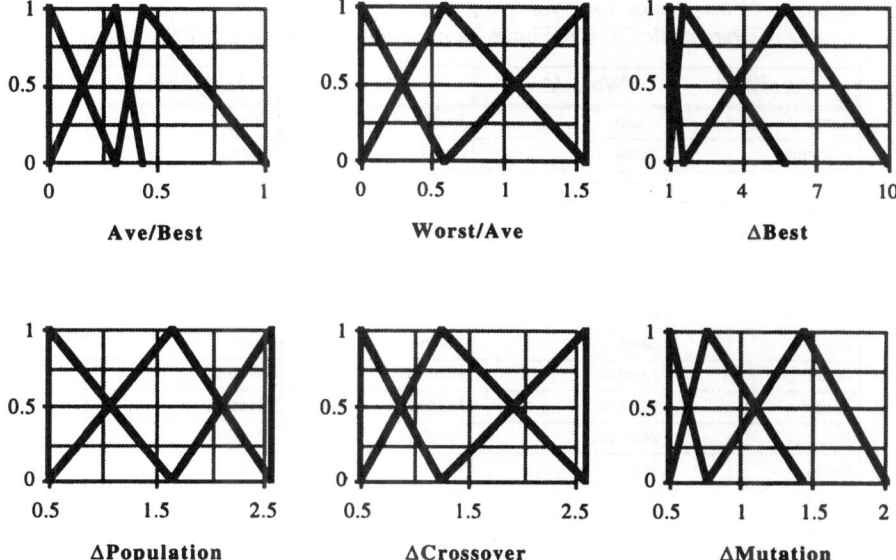

Fig. 3. Obtained membership functions for the optimized online fuzzy rule-base. **Ave/Best** represents the ratio between the average fitness and the fitness of the best individual's fitness. **Worst/Ave** represents the ratio between the worst individual's fitness and the average fitness. **ΔBest** represents the amount of change in the best fitness since the last control action. The control actions are **ΔPopulation**, **ΔCrossover**, and **ΔMutation** which control the change in population size, crossover rate, and mutation rate respectively. Each fuzzy set within an input variable is identified from left to right as Small, Medium, or Big. Each fuzzy set within an output variable is identified from left to right as Negative, None, or Positive. See Tables 2, 3, and 4 for rules.

After evaluating 665 fuzzy systems, the meta-level genetic algorithm produced a fuzzy system with the following initial conditions:

Initial Population Size:	10
Initial Crossover Rate:	0.942647
Initial Mutation Rate:	0.009903

The membership function definitions and rules are given in Figure 3 and Tables 2, 3, and 4. The first three columns represent the antecedent part and the fourth column is the consequent. For example, the first entry in the table should be read: **IF** Ave/Best is Small **AND** Worst/Ave is Small **AND** ΔBest is Small **THEN** change in population size is None.

According to the aggregate online fitness measure defined by (3), this Dynamic Para-

Table 2: Fuzzy rules for controlling population size in online Dynamic Parametric GA (see Figure 3 for membership function definition).

Ave/Best	Worst/Ave	ΔBest	ΔPopulation
Small	Small	Small	None
Medium	Small	Small	Negative
Small	Medium	Small	Positive
Small	Big	Small	Positive
Medium	Big	Small	None
Small	Small	Medium	Positive
Medium	Small	Medium	Negative
Big	Small	Medium	Positive
Small	Medium	Medium	Positive
Medium	Medium	Medium	Negative
Big	Medium	Medium	Negative
Small	Big	Medium	Negative
Medium	Big	Medium	Positive
Small	Small	Big	Positive
Small	Medium	Big	None
Medium	Small	Big	None
Big	Small	Big	Positive
Medium	Medium	Big	None
Big	Medium	Big	Positive
Small	Big	Big	None
Medium	Big	Big	Positive
Big	Big	Big	Positive

Table 3: Fuzzy rules for controlling crossover rate in online Dynamic Parametric GA (see Figure 3 for membership function definition).

Ave/Best	Worst/Ave	ΔBest	ΔCrossover
Small	Small	Small	Positive
Medium	Small	Small	Positive
Big	Small	Small	Positive
Small	Medium	Small	Negative
Medium	Medium	Small	None
Medium	Big	Small	Negative
Small	Small	Medium	None
Small	Medium	Medium	Negative
Medium	Medium	Medium	Negative
Big	Medium	Medium	Negative
Small	Big	Medium	None
Medium	Big	Medium	None
Big	Big	Medium	None
Small	Small	Big	Positive
Medium	Small	Big	Negative

Table 3: Fuzzy rules for controlling crossover rate in online Dynamic Parametric GA (see Figure 3 for membership function definition).

Ave/Best	Worst/Ave	ΔBest	ΔCrossover
Medium	Big	Big	Positive
Big	Big	Big	Positive

Table 4: Fuzzy rules for controlling mutation rate in online Dynamic Parametric GA (see Figure 3 for membership function definition).

Ave/Best	Worst/Ave	ΔBest	ΔMutation
Small	Small	Small	Negative
Big	Small	Small	None
Medium	Medium	Small	Negative
Small	Big	Small	Negative
Medium	Big	Small	Negative
Big	Big	Small	None
Small	Small	Medium	Positive
Medium	Small	Medium	None
Big	Small	Medium	Positive
Medium	Medium	Medium	Negative
Big	Medium	Medium	Negative
Small	Big	Medium	Positive
Medium	Big	Medium	None
Big	Big	Medium	None
Small	Medium	Big	Negative
Small	Small	Big	None
Medium	Small	Big	Positive
Medium	Medium	Big	None
Medium	Big	Big	Negative
Big	Big	Big	None

metric GA exhibited a 380% increase in performance over the optimized static genetic algorithm (calculated by dividing the Dynamic Parametric GA performance by the optimized static genetic algorithm performance) (see Table 5). Data for each of the

Table 5: Online performance measures for optimized static GA (OSGA) and Dynamic Parametric GA (DPGA).

Function	OSGA	DPGA
1	1069.525635	7555.996094
2	1479.180542	7952.011230
3	888.739319	1463.375610
4	579.341919	565.375854
5	833.273376	905.430664

122

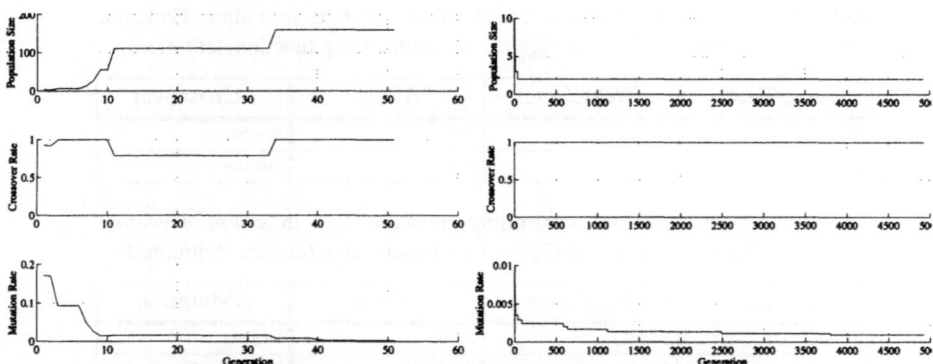

Fig. 4. Online performance of the Dynamic Parametric GA (DPGA) optimized for online performance on De Jong Function 3. Also shown are the online performance measures of the optimized static GA (OSGA online), simple static GA (SSGA), and random search (see Table 1 for GA search parameters). The plots on the right show the dynamic control of the population size, crossover rate, and mutation rate for a typical DPGA run on De Jong Function 3.

entries represent values averaged over ten independent runs starting from different initial conditions. The initial population of fuzzy systems and initial conditions used for determining the good online performance was seeded with an individual with static settings as prescribed by 'OSGA online' given in Table 1 (the system had no active rules and initial conditions set as prescribed by 'OSGA online'). For each fuzzy system produced for the Dynamic Parametric GA, the generation gap, window size, and selection strategy were fixed at 1.0, 1, and Elitist. After evaluating 665 fuzzy systems, the meta-level genetic algorithm produced a fuzzy system with 59 rules and the following initial conditions[14]:

Initial Population Size:	10
Initial Crossover Rate:	0.942647
Initial Mutation Rate:	0.009903

The left plot of Figure 4 shows the online performance vs. evaluations for the DPGA, OSGA online, SSGA, and random search for De Jong Function 3. The data in the figure is averaged from running each GA with ten different initial conditions.

The plots on the right show the dynamic control of the population sizing, crossover rate, and the mutation rate for a typical run on De Jong Function 3. Both the population size and mutation rate decrease toward the minimum value while the crossover rate remains high. The strategy that this particular DPGA has chosen is a conservative approach. Because the elite selection strategy is enabled and the population size goes to two, the search becomes a greedy hill climber. A good solution is not abandoned until a better one is found. In addition, the low mutation rate keeps the exploration relatively local.

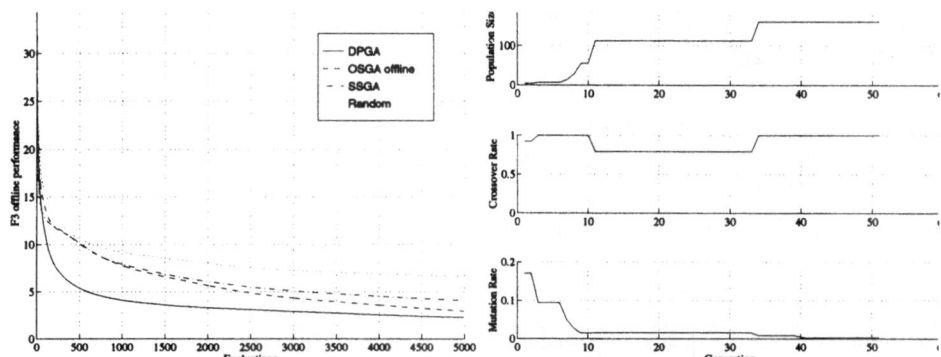

Fig. 5. Offline performance of the Dynamic Parametric GA (DPGA) optimized for offline performance on De Jong Function 3. Also shown are the offline performance measures of the optimized static GA (OSGA offline), simple static GA (SSGA), and random search (see Table 1 for GA search parameters). The plots on the right show the dynamic control of the population size, crossover rate, and mutation rate for a typical DPGA run on De Jong Function 3.

5.2 Dynamic Parametric GA for Offline Performance

As with the online search, the initial population of fuzzy systems and initial conditions used for determining the good offline performance was seeded with an individual identical with static settings, i.e. no rules, as prescribed by 'OSGA offline' given in Table 1. For each fuzzy system produced for the Dynamic Parametric GA, the generation gap, window size, and selection strategy were fixed at 0.9, 1, and Pure. After evaluating 373 fuzzy systems, the meta-level genetic algorithm produced a fuzzy system with 68 rules and the following initial conditions[14]:

Initial Population Size:	4
Initial Crossover Rate:	0.922059
Initial Mutation Rate:	0.170671

Figure 5 shows the offline performance vs. evaluations for the DPGA, OSGA offline, SSGA, and random search for De Jong Function 3 (as with the online data, this data is averaged over ten runs). The plots on the right show the dynamic control of the population sizing, crossover rate, and the mutation rate. As in the online control strategy, the mutation rate decrease toward the minimum value while the crossover rate remains high. However, the population size increases toward the maximum value. As the number of evaluations increases, random search becomes more difficult to out-perform. Although we expected the mutation rate to increase over time (a move toward random search behavior) we found that the control strategy relied more on the crossover operator than the mutation as it continued its search.

6 The Interaction Effects of Dynamic Crossover, Mutation, and Population Sizing

In this section, we present experimental results on different combinations of dynamic control parameters on offline performance. We compare eight combinations of results as shown in Table 6.

Table 6: Experimental combinations of dynamic parameters.

control parameters	experiment identifier						
	sss	dss	sds	ssd	dds	sdd	ddd
population	static	dynamic	static	static	dynamic	static	dynamic
crossover	static	static	dynamic	static	dynamic	dynamic	dynamic
mutation	static	static	static	dynamic	static	dynamic	dynamic

The experiment identifier reflects which control parameters were allowed to change and which fixed for a particular algorithm. These experimental results were generated by disabling the appropriate dynamic control of the DPGA. Figure 6 compares the offline performance for Function 3 using each of the combinations averaged over 20 different runs. The observed performance behaviors separated into three categories: not so good, good, and best (figures (a), (b), and (c) in Figure 6 respectively). The figure shows that in the case of adding only dynamic population sizing, only a marginal improvement is made. Also, in general, modulating the crossover rate offers moderate improvements. In addition, it is usually the case that adding dynamic mutation always results in a performance gain.

7 Conclusions and Extensions

We have presented a framework for controlling genetic algorithm and demonstrated how to apply it to study the interdependencies of dynamic crossover, mutation, and population sizing. Our framework includes techniques to automatically design genetic algorithm control strategies according to a given search performance metric. Using these techniques, we have designed adaptive genetic algorithms, referred to as Dynamic Parametric Genetic Algorithms, with control strategies that show improved performance over simple static and optimized static GAs. Further comparisons with other GA methods are warranted.

Results from experiments aimed at isolating the effects of dynamic population sizing, mutation rate, and crossover rate shows that the dynamic mutation contributes most to high online and offline performance.

We would like to emphasize that the experimental results we report have been obtained for a specific instance of the class of Dynamic Parametric GAs; our technique can be applied to systems with other inputs and outputs. Research on rule compaction and determining relevant input variables should be explored. Also, more analysis needs to be performed on the resulting systems. In addition, obtaining search behaviors other than ones directed at high offline and online measures warrant investigation.

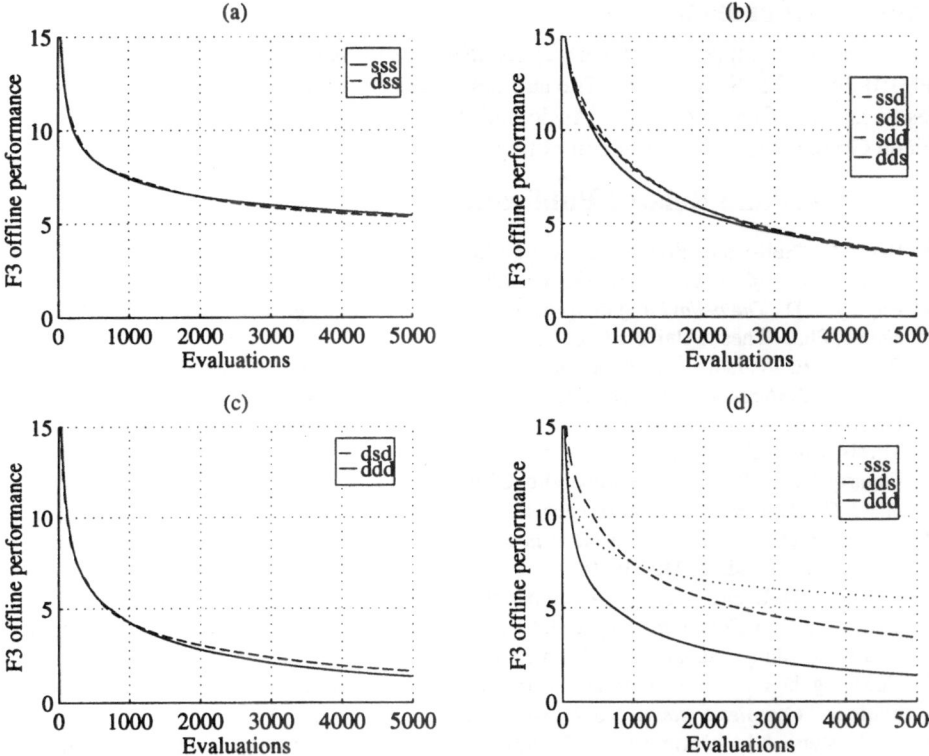

Fig. 6. Comparison plots of DPGAs with different dynamic control parameters. The performance characteristics fall into three categories. Each category is shown in plots (a), (b), and(c). Plot (d) groups selected GAs from each of the categories.

The framework presented in this paper can be extended and generalized to realize self-adapting learning algorithms based on other learning algorithms such as simulated annealing or gradient techniques. The underlying benefit that this technique offers is a method to tune and adapt a particular learning algorithm to the problem at hand. The initial cost of using our method to design an adaptive learning algorithm can be high, however, the procedure need not be performed frequently. The goal of our technique is to design a learning algorithm that can be repeatedly applied to a particular class of problems, such as fuzzy system design using a specific fuzzy system representation.

It is known that the efficiency and effectiveness of a particular learning algorithm hinge on:1) the representations they operate on, 2) the settings of the learning algorithm parameters, and 3) the error functions used to drive the learning. Although all instances of the same algorithm are conceptually the same, in practice, systematic differences are always present. Our method provides a technique that can minimize the effects of systematic deficiencies to obtain higher performance learning algorithms.

8 Acknowledgments

This research is supported in part by NASA Grant NCC-2-275, EPRI Agreement RP8010-34, and BISC program. The authors would also like to thank Professor David Wessel and the Center for New Music and Audio Technologies at UC Berkeley and Silicon Graphics Inc. for use of computing resources.

9 References and Related Publications

[1] Back, T., "Self-adaptation in genetic algorithms," *Proceedings of the First European Conference on Artificial Life*, Paris, France, 1991, pp.263-271.

[2] Bagley, J. D., *The behavior of adaptive systems which employ genetic and correlation algorithms*, Ph.D. Thesis, University of Michigan, 1967.

[3] Baker, J. E., "Adaptive selection method for genetic algorithms," *Proceedings of an International Conference on Genetic Algorithms* (ICGA'85), 1985, pp.101-111.

[4] De Jong, K. A., *An analysis of the behavior of a class of genetic adaptive systems*, Ph.D. Thesis, University of Michigan, 1975.

[5] Fogarty, T. C., "Varying the probability of mutation in the genetic algorithm," *Proceedings of the Third International Conference on Genetic Algorithms* (ICGA'89), 1989, pp.104-109.

[6] Goldberg, D. E., *Genetic algorithms in search, optimization, and machine learning*, Addison-Wesley, Reading, MA, 1989.

[7] Goldberg, D. E. and Deb, K., "A comparative analysis of selection schemes used in genetic algorithms," in *Foundations of genetic algorithms* (FOGA'90), ed. Rawlins,G., Morgan Kaufmann, San Mateo, CA, 1991, pp.69-93.

[8] Goldberg, D. E., Deb, K. and Clark, J. H., "Genetic algorithms, noise, and the sizing of populations," *Complex Systems*, Vol.6, No.4, 1992, pp. 333-362.

[9] Grefenstette, J. J., "Optimization of control parameters for genetic algorithms," *IEEE Transactions on Systems, Man, and Cybernetics*, Vol.16, No.1, 1986, pp. 122-128.

[10] Hesser, J. and Manner, R., "Towards an optimal mutation probability for genetic algorithms," *Parallel Problem Solving from Nature. 1st Workshop, PPSN 1 Proceedings*, Dortmund, West Germany, 1990, pp.23-32.

[11] Holland, J. H., *Adaptation in Natural and Artificial Systems*, MIT Press, Cambridge, MA, 1975.

[12] Karr, C., Freeman, L. and Meredith, D., "Improved Fuzzy Process Control of Spacecraft Autonomous Rendezvous Using a Genetic Algorithm," *SPIE Intelligent Control and Adaptive systems*, 1989.

[13] Kosko, B., *Neural Networks and Fuzzy Systems*, Addison-Wesley, Englewood Cliffs, NJ, 1992.

[14] Lee, M.A., *Automatic Design and Adaptation of Fuzzy Systems and Genetic Algorithms using Soft Computing Techniques*, Ph.D. Thesis, University of California, Davis, 1994.

[15] Lee, M. A. and Takagi, H., "Embedding A Priori Knowledge into an Integrated Fuzzy System Design Method based on Genetic Algorithms," *Proc. of the IFSA Congress* (IFSA'93), Seoul, Korea, 1993, pp.1293-1296.

[16] Lee, M. A. and Takagi, H., "Integrating design stages of fuzzy systems using genetic algorithms," *Proc. IEEE Int. Conf. on Fuzzy Systems* (FUZZ-IEEE '93), San Francisco, CA, 1993, pp.612-617.

[17] Lee, M. A. and Takagi, H., "Dynamic Control of Genetic Algorithms using Fuzzy Logic Techniques," *Proceedings of the Fifth International Conference on Genetic Algorithms* (ICGA'93), Morgan-Kaufmann, San Mateo, CA, 1993, pp.76-83.

[18] Schaffer, J. D. and Morishima, A., "An adaptive crossover distribution mechanism for genetic algorithms," *Proceedings of the Second International Conference on Genetic Algorithms* (ICGA'87), MIT, Cambridge, MA, 1987, pp.36-40.

Unsupervised/Supervised Learning for RBF-Fuzzy System
-Adaptive Rules, Membership Functions and Hierarchical Structure by Genetic Algorithm-

Koji Shimojima, Yasuhisa Hasegawa and Toshio Fukuda

Department of Mechano-Informatics and Systems
Nagoya University
Furo-cho, Chikusa-ku, Nagoya 464-01, Japan
Tel: +81-52-789-2717, Fax:+81-52-789-3115
E-mail: koji@mein.nagoya-u.ac.jp

Abstract: Recently, fuzzy reasoning has been used in many fields and places. In order to apply the reasoning to the various fields, the tuning and optimizing method of the fuzzy reasoning is the key issue. Some self-tuning methods have been proposed so far. However, these conventional self-tuning methods do not have sufficient capability of learning. In this paper, we propose new unsupervised/supervised self-tuning methods for fuzzy reasoning, which consists of membership functions expressed by the radial basis function with an insensitive region. Learning is carried out by a genetic algorithm. The gradient decent method is also used for tuning the shapes and location of membership function and consequent parts in case of supervised learning. The effectiveness of the proposed methods is shown by some numerical examples.

1. Introduction

In recent years, fuzzy systems such as fuzzy reasoning, fuzzy modeling and fuzzy logic controllers have been used in many fields for engineering, medical engineering and social science. Fuzzy control systems have already been deployed in home appliances, transportation systems and manufacturing systems. We also have been studying(Fukuda *et al.*, 1993, Shimojima *et al.*, 1992, 1993) to sensor integration systems applying fuzzy inference.

Fuzzy systems can represent human knowledge or experiences as fuzzy rules. However, fuzzy systems have some problems. In the most fuzzy systems, the shape of membership functions of the antecedent, the consequent and fuzzy rules were determined and tuned through trial and error by human operators. Therefore it takes many iterations to determine and tune them, and it is very difficult to design the optimal fuzzy system. This problem is very serious, in case of complex systems.

In order to solve this problem, some self-tuning methods have been proposed such as Fuzzy Neural Network(Horikawa *et al.*, 1992, Higgins *et al.*, 1992) that is back propagation algorithm(Rumelhart *et al.*, 1986) applied to the learning, fuzzy learning controller applying the radial basis function(Katayama *et al.*, 1993, Linkens *et al.*, 1993), using genetic algorithms for deciding the shapes of membership functions and fuzzy rules(Whitley *et al.*, 1990, Nomura *et al.*, 1991, Lee *et al.*, 1993) and so on (Nomura *et al.*, 1991).

These methods can learn faster than neural networks. However, the operator must determine the number and shapes of membership functions before learning, and the learning ability and accuracy of approximation are related to the number or shapes of membership functions. Fuzzy inference with many membership functions and fuzzy rules has a high learning capability, however redundant rules or unlearned rules may be existed. In complete rule sets, the number of rules is the product of the number of membership functions for each input, and the number of rules is increased as exponential with input dimension. Therefore, the operators must carefully decide the structure of the membership functions. For this problem, the hierarchical fuzzy inference has been proposed to reduce the number of fuzzy rules. However, this method also has a major problem. The operator must design a hierarchical structure of the fuzzy reasoning considering the relationship among input and output variables because input variables are classified into some groups, with which local outputs in the hierarchical structure are determined.

Fuzzy inference based on RBF networks that add a new rule for the maximal error point through the learning process has been proposed. In this method, fuzzy rules depend on the learning data set and if the learning data is biased, there are some unlearning area or redundant fuzzy rules, therefore the learned fuzzy rules are not optimized. This method also has the increasing fuzzy rule problem, because this method does not integrate or delete fuzzy rules, only adds a new fuzzy rule. The addition of fuzzy rules increases calculation time and memory consuming.

In this paper, we propose a new type of self-tuning fuzzy inference. The membership function of the antecedent is expressed by the radial basis function with an insensitive range. Both supervised/unsupervised learning algorithms are used and based on the genetic algorithm. The supervised learning also uses a gradient decent method to tune the shapes of membership functions and the consequent values. We also propose supervised learning for a hierarchical fuzzy inference. The hierarchical structure is very powerful for increasing of input values. The proposed hierarchical fuzzy inference considers the relationship among input and output variables, and it does not lose the accuracy of the output values with fewer fuzzy rules.

In the following sections, we explain how to apply the genetic algorithm to the learning algorithm and how to organize the hierarchical structure by the genetic algorithm. We apply our methods to a function approximation problem and to the problem of following trajectory to show the effectiveness of these learning methods.

2. Unsupervised/Supervised Learning for Fuzzy Reasoning

2.1 Fuzzy Inference based on RBF

The shape of the membership function is the radial basis function with an insensitive range c that is useful for reducing the number of the membership functions. The membership function in the i-th input variable and the j-th fuzzy rule is expressed by

$$f_{ij}(x_i) = \begin{cases} 0, & \text{if } |x_i - a_{ij}| \le c_{ij} \\ |x_i - a_{ij}| - c_{ij}, & \text{if } |x_i - a_{ij}| > c_{ij} \end{cases}, \tag{1}$$

$$\mu_{ij} = \exp\left(-b_{ij} \cdot f_{ij}(x_i)^2\right), \tag{2}$$

where *a, b, c* are the coefficients that decide the shape of membership functions shown in Fig. 1, and *i* is the input number, *j* is the fuzzy rule number.

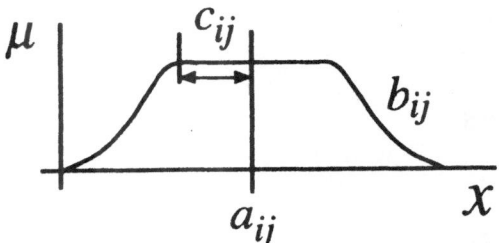

Fig. 1. Membership Function based on RBF

We describe the calculating formulas between input values and output values. The fitness value μ_j of the rules and the output value Yp are expressed by

$$\mu_j = \prod_{i \in I} \mu_{ij} \ , \tag{3}$$

$$Y_p = \frac{\sum_{j=1}^{J} \mu_j \cdot w_j}{\sum_{j=1}^{J} \mu_j} \ , \tag{4}$$

where *p* is the data set's number.

2.2 Coding

To encode the information of the membership functions, we use 3n+1 bits for each membership function: each coefficient, *a, b,* and *c,* uses n bits and 1 bit is used to show rule validity. In case of unsupervised learning, m bits are used to express the consequent value. The following equations (5), (6), (7), and (8) are used to translate the chromosome into the membership function parameters.

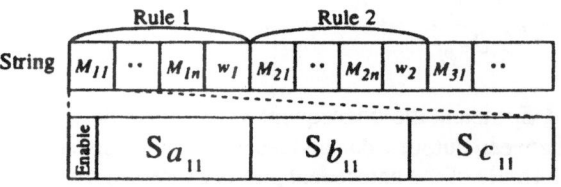

Fig. 2. Coding

$$a_{ij} = \frac{2Sa_{ij}}{2^n - 1} - 0.5, \tag{5}$$

$$b_{ij} = 100\left(\frac{Sb_{ij}}{2^n - 1}\right)^3, \tag{6}$$

$$c_{ij} = \frac{Sc_{ij}}{2^n - 1}, \tag{7}$$

$$W_i = \frac{2w_i}{2^m - 1} - 0.5 \tag{8}$$

2.3 Learning Methods
2.3.1 Supervised Learning

Several papers have proposed automatic design(self-tuning) methods. Much of the work has focused on tuning membership functions. For example, neural networks are used as membership value generators and fuzzy systems are treated as networks. Back-propagation techniques are used to adjust the shapes of membership functions. However, these tuning methods have a weak point: they often converge to a local minimum, because the convergence of tuning depends on the initial conditions such as the number and shapes of membership functions.

On the other hand, another method using a Genetic Algorithm(GA) is proposed for the purpose of auto-tuning and optimization of the structure of the fuzzy model. The GA is an optimization method that uses a stochastic search algorithm based on the biological evolution process. However, the GA search is coarse and inefficient at finding the optimal value.

We propose a fuzzy reasoning method based on radial basis function (RBF) networks and two supervised learning methods. One tuning method is the coarse tuning with GA, and the other is the fine tuning by the gradient descent method. We can escape from the local minimum by using GA, and we also can get the optimal fuzzy reasoning by the gradient descent method. Therefore, we use both GA and the gradient descent method alternately, as shown in Fig. 3. Furthermore, GA is able to add and remove the membership functions and rules through the learning process. The proposed method can not only tune the shapes of membership functions and the consequent value, but also adjust the number of membership functions and rules.

The tuning formula of the consequent part is expressed by

$$w_{pj} = w'_{pj} - k_w \frac{\partial E_p}{\partial w_{pj}}, \tag{9}$$

where w, p, k_w and E_p means the consequent value, the rule number, coefficient of learning and error between output value and teacher signal respectively.

The tuning formula of the antecedent part is expressed by

$$a_{ij} = a'_{ij} - k_a \frac{\partial E_p}{\partial a_{ij}}, \tag{10}$$

$$b_{ij} = b'_{ij} - k_b \frac{\partial E_p}{\partial b_{ij}}, \tag{11}$$

$$c_{ij} = c'_{ij} - k_c \frac{\partial E_p}{\partial c_{ij}} ,$$
(12)

where k_a, k_b and k_c are the learning coefficients.

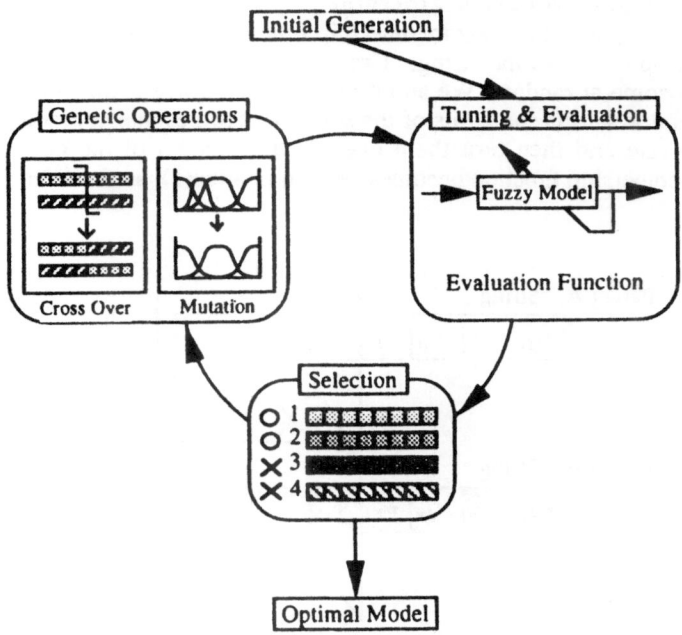

Fig. 3.   Outline of Tuning Method

2.3.2 Unsupervised Learning

Unsupervised learning is very useful in many fields, because designers; (1) do not have to make the teach signals which are acquired by some mathematical methods, and (2) often can not acquire the mathematical model of the environment, because of non-linear property. It is, however, the much more difficult than the supervised learning, i.e., it takes much time to obtain the desired fuzzy systems.

In case of unsupervised learning, only GA is utilized for adding, deleting and tuning of the membership functions and the consequent values.

2.4 Selection

We rank the each strings based on fitness value F expressed in eq. (13). We employ the elite preserving strategy to keep the highest fitness chromosomes and random selection to prevent initial convergence.

$$F = \alpha_1 A + \alpha_2 M + \alpha_3 L + \alpha_4 R ,$$
(13)

where A, M, L, R and α represent the mean square error between teacher signals and output values, the maximal square error between teacher signal and output value, the number of the membership function, the number of the rules, and the coefficients respectively.

2.5 Crossover and Mutation Operator

In order to generate a new membership function and rule, we apply crossover and mutation operators to the strings. Crossover and mutation operators select the target chromosome at random. We adopt two point crossover, shown in Fig. 4. The mutation operator selects some bits of the selected chromosome at random based on the mutation rate and then turn them over. If the first bit of the chromosome is chosen, its membership function becomes dead(0) or alive(1), as shown in Fig. 5.

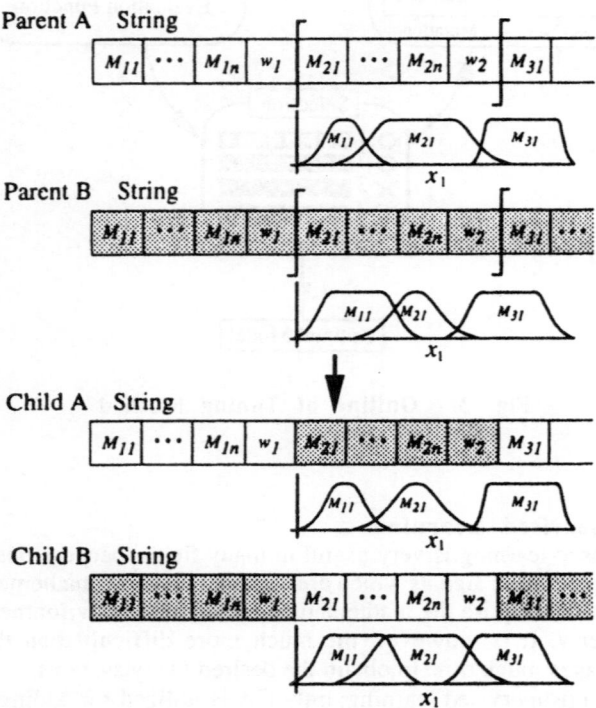

Fig. 4. Crossover

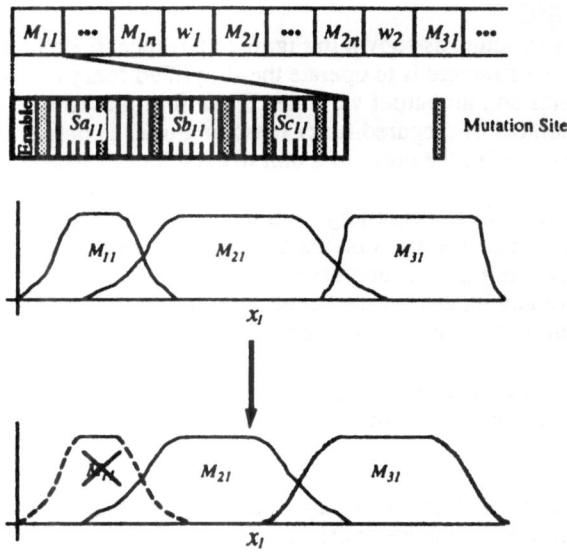

Fig. 5.   Mutation

3. HIERARCHICAL STRUCTURE OF FUZZY MODEL

3.1 Feature of Hierarchical Structure

Hierarchical structure is very effective to reduce the number of fuzzy rules, because each fuzzy reasoning unit of the hierarchical structure just has to reason with a few input variables classified into some groups. However the input variables should be classified into some groups as they correspond to a relationship among the input and output variables in teaching data. Some input variables which have the closer relationship than the others should be put into the same fuzzy reasoning unit. It is very difficult to find the relationship from numerical teaching data of nonlinear system. In practice, the structure has been determined by designers only when the relationship can be known.

The following method using the GA is proposed for the purpose of self-organization of the hierarchical structure. The optimal hierarchical structure of fuzzy reasoning is able to realize a higher precision with fewer rules than sub-optimal structures. To choose the optimal hierarchical structure, we use an evaluation function as a strategy that adopts a system with fewer rules and more accurate outputs.

3.2 Hierarchical Structure Organization Algorithm

The algorithm to make the fuzzy model with hierarchical structure is composed of two main process: Organizing the hierarchical structure and the tuning fuzzy model. The latter process to tune fuzzy model is carried out only when the fuzzy model with sufficient precision and permissible size of fuzzy model is not acquired in the former process. In this process, we use the supervised learning explained in section 2.3.1 The former process to organize the hierarchical structure of the fuzzy system is following;

INITIAL SETTING

1 Prepare a tree structure as shown in Fig. 6, in advance. Each unit that constructs the hierarchical structure is to operate the simplified fuzzy reasoning with a few input variables and an output variable. These formulas are (1),(2),(3),(4). The maximum number of prepared fuzzy reasoning units is defined by the function U in eq. (14). Assign the order of a unit to each input variable at random. (Fig. 6)

2 Check all units of fuzzy reasoning from the bottom layer. If the unit has two or more inputs, it reasons and outputs to the connected upper layer unit. If the unit has only one input, it outputs the same value as input value. If the unit does not have any inputs, it is removed from the structure. (Fig. 6)

3 Set some membership functions so that they divide an input space equally. The initial values of consequent parts of all units are zero.

4 Operations-(1), (2), and (3) are repeated until an initial generation can be set up. A hierarchical structure of fuzzy system is encoded into a string as the genetic information.

TUNING

5 Calculate output values of fuzzy reasoning units from the bottom layer. Besides calculate the square errors between output values of top layer unit and desirable data as teaching signal. The function is defined in eq. (15).

6 Get the errors of each reasoning unit from the error of the top layer unit, using back-propagation technique. The function is defined in eq. (16).

7 Tune the only consequent values in all units by using the gradient method, in eq. (9). The membership functions of the antecedent part are not tuned. The tuning will complete in case when the fitness value converges or tuning time reaches pre-determined learning times.

SELECTION

8 Evaluate each individual based on the fitness function, eq. (13). The fitness function consists of the number of the membership functions and rules of all units, and of the mean square error and of maximum square error between output values and desirable ones of only the top unit. This process will complete when the best fitness value becomes less than the target one or searching time reaches the limited generation times. Otherwise go to next procedure.

9 Select the individuals based on the fitness value and at random, to the next generation.

CROSSOVER and MUTATION

10 Operate crossover and mutation. These operations create new strings, shown in Fig. 8. The flow goes back to procedure (5).

$$U = 2^{i-2} \ , \tag{14}$$

$$E_{1n} = \frac{1}{2}\left(Y_{1n} - Y_{1n}{}^{*}\right)^{2}, \tag{15}$$

$$E_{pn} = \frac{\partial E_{qn}}{\partial Y_{pn}} \ , \tag{16}$$

where

Eln: summation of squared error of n-th data in the top unit,
Epn: summation of squared error of n-th data in p-th unit,
Ypn: output value of the n-th data in p-th unit,
Ypn*: teaching signal for *Ypn*,
n: the number of data,
p: the order of the fuzzy unit,
q: the order of the upper fuzzy unit connected to p-th fuzzy unit.

3.3 Coding

Each string has information as to where each input variable should be input. For example input variable No. 1 shown in Fig. 6 is input to unit No. 16.

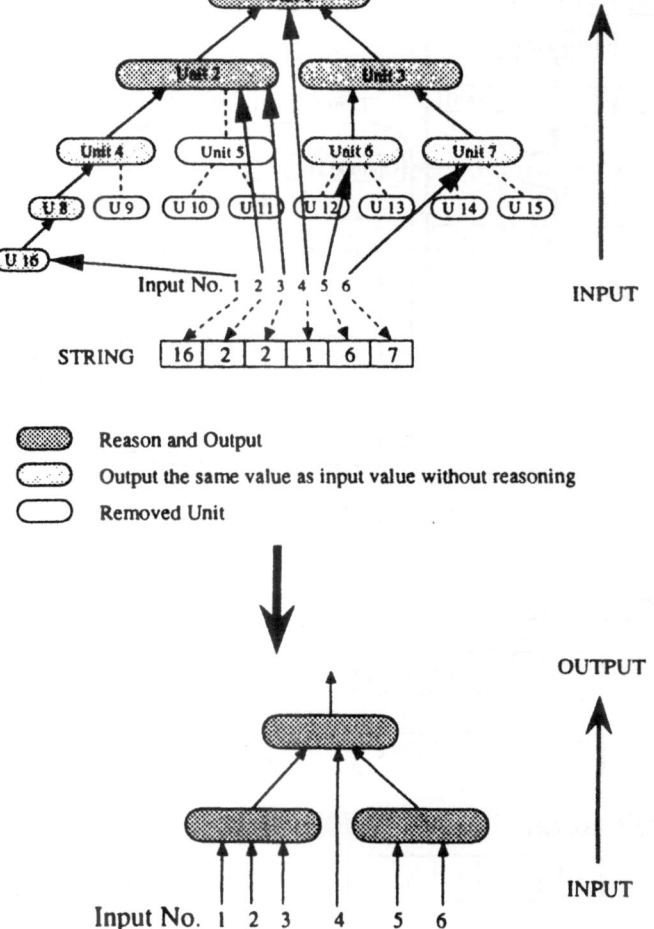

Fig. 6. Hierarchical Structure and Coding Method
The lower structure is simplified from the upper one.

3.4 Selection

To choose the fuzzy models with the optimal hierarchical structure, we use a fitness function as a strategy that adopts a system with fewer rules and more accurate outputs. The fitness function is the same eq. (13). We rank the tuned strings based on fitness value F. We employ the elitist preserving strategy to keep higher fitness gene and the random selection to prevent the initial convergence.

3.5 Crossover and Mutation Operator

In order to generate new hierarchical structures, we apply the crossover and mutation operations. The few fittest individuals and several individuals selected at random are used in crossover and mutation operations. We adopt multi-point crossover and the points are selected at random, shown in Fig. 8. In mutation operation, a part of gene selected at random is exchange to possible value determined at random.

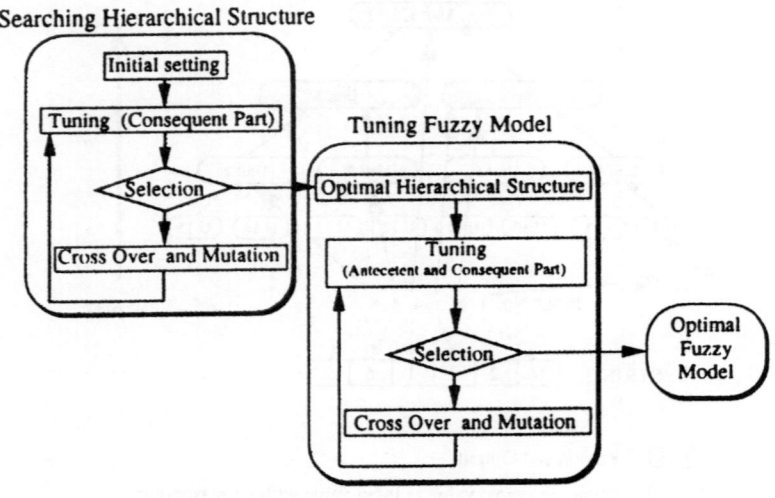

Fig. 7. Flow Chart

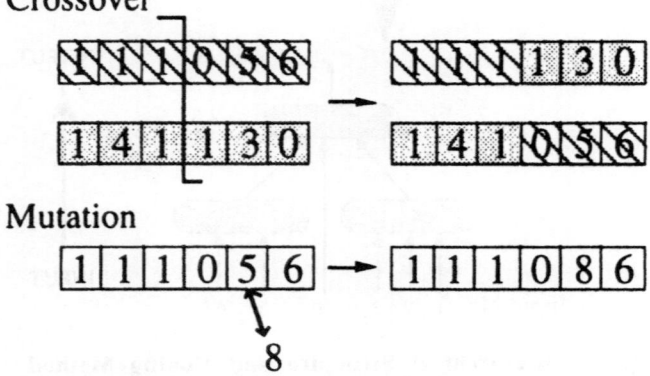

Fig. 8. Crossover and Mutation

4. Simulation Results

4.1 Unsupervised/Supervised Learning

In order to show the effectiveness of the proposed fuzzy reasoning with unsupervised/supervised learning, we apply fuzzy reasoning to approximate the function expressed eq. (17). The maximal number of rules and membership functions are 20 and 40. Population size is 60 (10 individuals from the parents + 20 from the crossover operation + 20 from mutation operation + 10 tuned by BP) in case of the supervised learning, and 50(10 from parents + Cross over 20 + Mutation 20) in case of the unsupervised learning. Mutation rate is 3%. Generation is 1000. The number of learning data set is 9x9 points.

$$y = 1, \qquad\qquad\qquad x_1 < 0.5 \qquad\qquad (17)$$
$$y = 1 - 4x_2(1 - (x_1 - 0.5)^2), \quad otherwise$$

Figs. 9 to 11 shows the result of supervised learning. The prefable fuzzy model has the least number of membership functions and rules. Therefore, parameters of the fitness function are set as follows; $\alpha_1 = 1.0$, $\alpha_2 = 1.0$, $\alpha_3 = 0.01$, and $a_4 = 0.01$. In order to reduce the number of the membership functions and fuzzy rules, the weight for them are large. In the fitness function, a membership function is considered as the same worth of 0.01 of the average error or the maxmam error. Therefore the maximam or average error would be converged around 0.01 in this simulation. The learning coefficients are kw=0.3, ka=0.2, kb=0.001 and kc=0.001.

Fig. 9 shows that the maxmam error is almost converged under 0.01 and the number of membership functions is decreased through the generations. After the learning, the number of membership functions is three, the number of rules is two, average error is 8.38×10^{-4} and maximum error is 6.13×10^{-3}. Fig. 14 shows the contents of membership functions and the consequent values. The approximated function is expressed by only 3 membership functions with two rules.

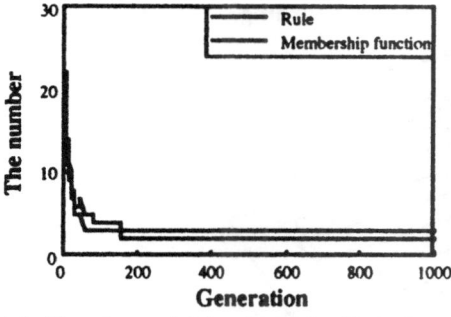

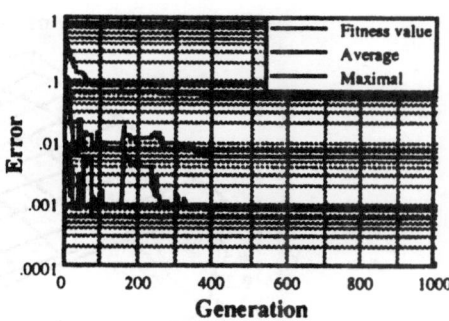

(a) Number of Membership Functions (b) Fitness values and Errors vs.
and Fuzzy Rules vs. Generation   Generations

Fig. 9.   Learning results of supervised learning

Figs 12 and 13 shows the results of unsupervised learning. The number of membership function is six, the number of rules is four, average error is 4.26×10^{-3}, and maximum error is 2.36×10^{-2}.

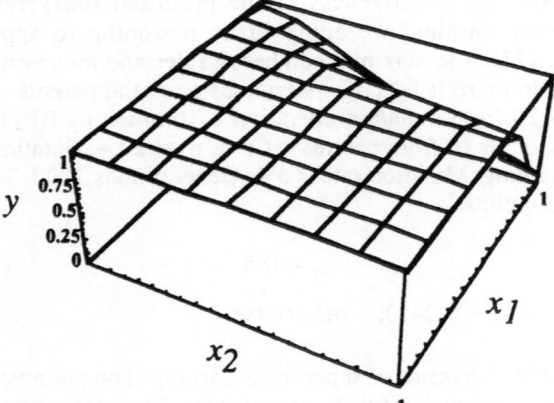

Fig. 10. Result of Supervised Learning

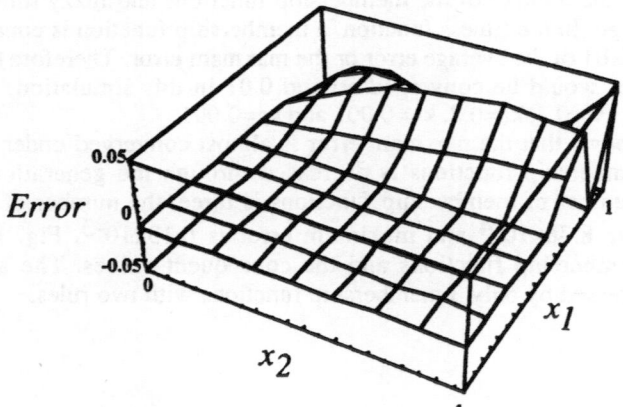

Fig. 11. Error of Supervised Learning

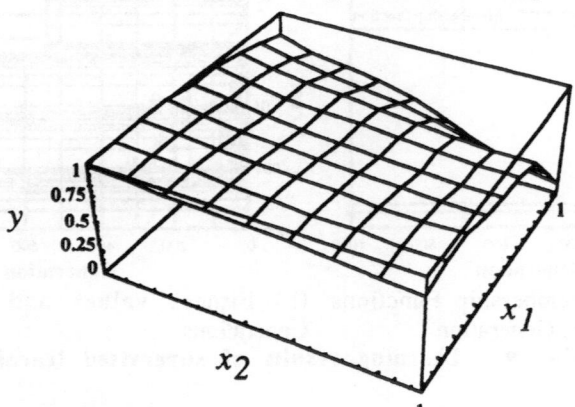

Fig. 12. Result of Unsupervised Learning

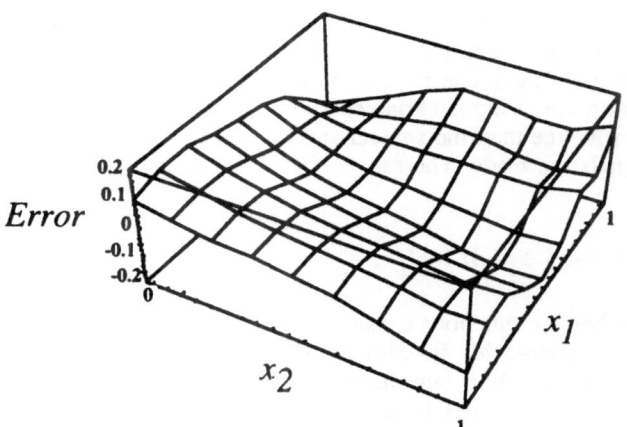

Fig. 13.   Error of Unsupervised Learning

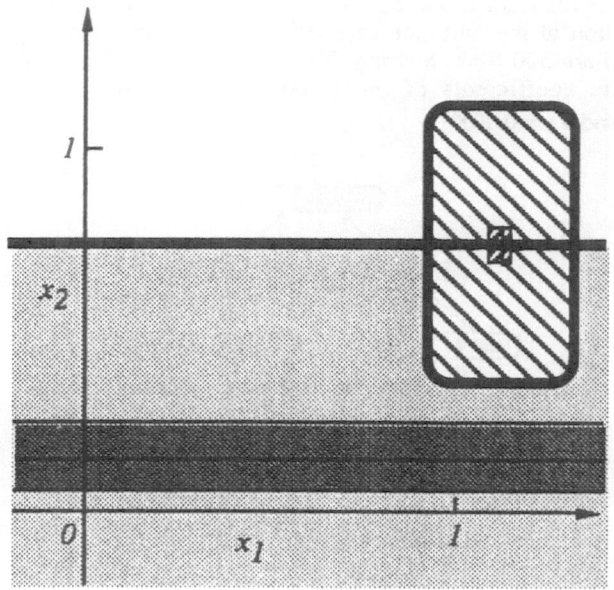

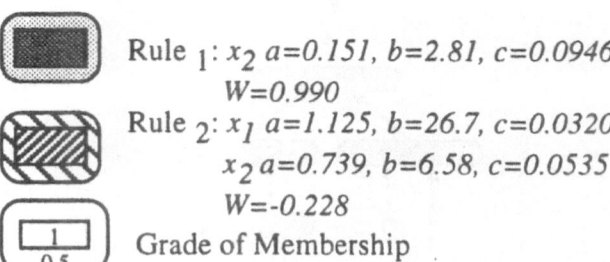

Rule $_1$: x_2 $a=0.151$, $b=2.81$, $c=0.0946$
            $W=0.990$
Rule $_2$: x_1 $a=1.125$, $b=26.7$, $c=0.0320$
            x_2 $a=0.739$, $b=6.58$, $c=0.0535$
            $W=-0.228$
Grade of Membership

Fig. 14.   Contents of Fuzzy Rules (supervised learning)

4.2 Hierarchical Structure

The proposed method is applied to multi-input function approximation problems in order to show the effectiveness. We consider a nonlinear system shown in eq. (18). This equation has 5 input variables and 1 output. In this equation, the first term has more influence than the second and the third term on output values. 100 sets of teach signals are made. The range of each input and output variables takes from 0 to 1.

$$Y = (4X_1X_2 + X_3X_4 + X_5)/6 \qquad (18)$$

The three membership functions are set so that they divide an input space equally. Each rule has each membership function but owns them jointly. For example, if the unit of fuzzy reasoning has two inputs, it has 9(=3x3) rules and 18(=9x3) membership functions. If three inputs, it has 27(=3x3x3) rules and 81(=27x3) membership functions. The initial values of the consequence part are 0.

We search the optimal hierarchical structure using the GA. One generation consists of 20 strings: 4 strings are generated from the top strings of the last generation. 16 strings are made by the cross over operations and operated the mutation operation at the mutation rate 10%. The maximum tuning times of the gradient method are 500 times a string. The learning coefficient of the consequent part is 0.3. The coefficients of the fitness function are $\alpha_1 = 1.0$, $\alpha_2 = 0.1$, $\alpha_3 = 0.00001$, and $\alpha_4 = 0.00001$.

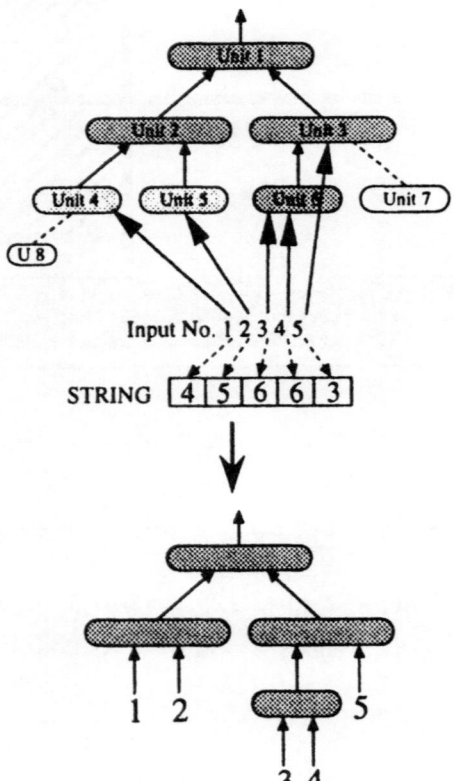

Fig. 15. The Optimal Structure Obtained by the Proposed Method

Table 1 Comparison of Single/Hierarchical Structure:
It represents the results of fuzzy models with the two different structures. The hierarchical fuzzy model could realize the same precision with less than half of 78 rules in fuzzy model with single layer.

	Single Structure	Hierarchical Structure
Max. Error	2.76×10^{-3}	1.92×10^{-3}
Average Error	5.63×10^{-4}	1.81×10^{-4}
No. of Rules	78	36
No. of Membership functions	157	64

Seven times are tried with various random seeds, and the mean generation required to obtain the optimal hierarchical structure shown in Fig. 15 is 4.2 generations. The mean generation is 6.2 with simple GA. This fuzzy model has 36 rules and 72 membership functions. Mean square error is 3.92×10^{-4}. Max. square error is 2.75×10^{-3}. The obtained structure represents the proper relationship among input and output variables in the equation.: Two input variables which are close each other, for example No. 1 and No. 2, are input to the same unit, amd the more influent term to the output value are located at the closer unit to output unit. Therefore we can say we obtained the suitable hierarchical fuzzy model with this method.

We made a fuzzy model with one layer from the same teaching data in order to compare the number of rules in it with that of rules in the hierarchical fuzzy model. In this case the antecedent and consequent parts in both fuzzy models are tuned using GA and the gradient method, which is explained in the section 2.3.1. The learning coefficient of consequent part is 0.1, and that of antecedent parts are k_a, k_b, k_c=0.001. The table 1 shows that the hierarchical fuzzy model could realize the same precision with less than half of 78 rules in fuzzy model with single layer.

4.3 Following Trajectory by Unsupervised Learning

In this subsection, we applied our unsupervised learning RBF fuzzy inference for the control system. In case of the control system, the unsupervised learning system is very important since some systems would not be analyzed in precise and the mathematical model of the controlled object based on the analysis has some difference with the real system. Therefore, some cases are very difficult to adjust the control system as the sufficient performance and it takes many trials to make the control system by adjusting many parameters of the controller.

We make the fuzzy controller perform position control of a single degree-of-freedom, spring-mass-damper system with variable friction. The equation of the motion for the cart(mass) in the system shown in Fig. 16 is expressed as

$$m\ddot{x} + c\dot{x} + kx^3 + \mu(x,\dot{x}) = f, \tag{19}$$

where, x, xd, m, c, k, and μ means the position of the cart, the desired trajectory of the cart, the mass of the cart connected of the spring and the damper, the damping coefficient, the nonlinear spring constant, and the friction coefficient that varies on the position and the velocity of the mass shown in Fig. 18 and 19, respectively. The desired trajectory of the cart is expressed as follows:

$$
\begin{aligned}
x_d &= 0.5 - \frac{1}{2}\cos(\pi t) & (0 \leq t \leq 1) \\
x_d &= 1 & (1 \leq t \leq 2, 5 \leq t \leq 6) \\
x_d &= 0.5 - \frac{1}{2}\cos(\pi(t-1)) & (2 \leq t \leq 3) \\
x_d &= 0 & (3 \leq t \leq 4) \\
x_d &= 0.5 - \frac{1}{2}\cos(\pi(t-4)) & (4 \leq t \leq 5)
\end{aligned}
\tag{20}
$$

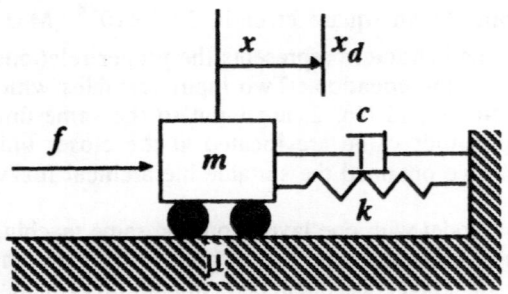

Fig. 16.   The Controlled Object

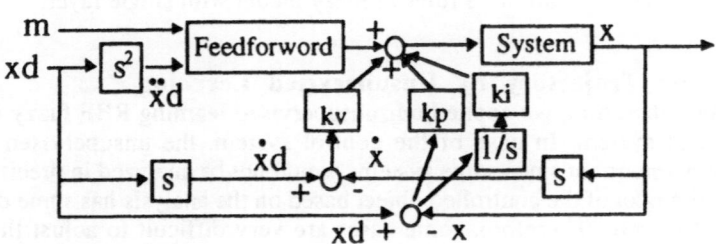

Fig. 17.   Block Diagram of the System

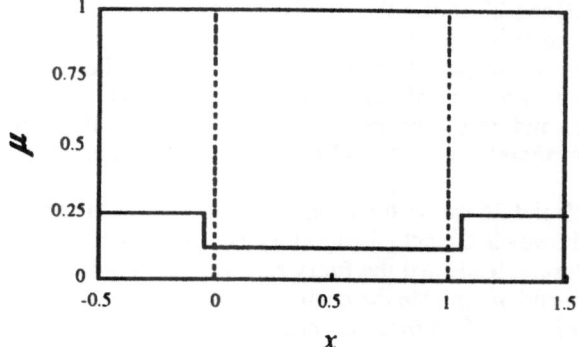

Fig. 18.   Dynamic Friction

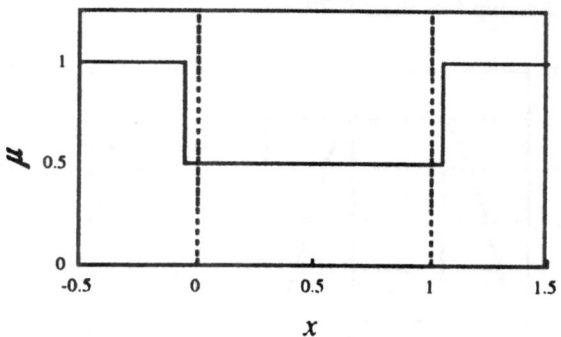

Fig. 19.   Static Friction

Inputs of the fuzzy controller are xd, $\dot{xd}$, $\ddot{xd}$, and m and output is f. The fitness function is

$$F = \alpha_1 \sum_{t=0}^{T/dt} (xd(t) - x(t))^2 + \alpha_2 Max(xd(t) - x(t))^2 + \alpha_3 L + \alpha_4 R, \qquad (21)$$

where T is the total controlling time, dt is the sampling interval.

The number of rules and membership functions are 40 and 160 (= 4X40). Population size is 50(Parents 10 + Cross over 20 + Mutation 20). Iteration time is 2000 generations. Parameters of GA are as follows: $\alpha_1 = 1.0$, $\alpha_2 = 60.0$, $\alpha_3 = 0.000001$, and $a_4 = 0.000001$. Parameters of PID controller, the mass, the desired trajectory are as follows;

xd	0 to 1.0 m
m	0.5kg, 1.0kg, and 0.75 kg(unlearned object)
k	200.0 N/m,
c	0.5 Ns/m ($\dot{x}>0.0$), 2.0Ns/m($\dot{x}<0.0$),
kp	100.0
kv	10
ki	5
T	6 sec.
dt	0.01 sec.

Figures 21 and 22 shows the simulation results in case of m=0.5kg. In case of PID controller, the actual trajectory is different from the desired trajectory because of the friction. However in case of the PID with fuzzy controller, this object has been learned and we get the good actual trajectory with small trajectory errors.

Figures 23 and 24 shows the results in case of m=1.0kg, this object is also the learning object therefore the actual trajectory is very similar with the desired trajectory.

Figures 25 and 26 shows the results of m=0.75kg. This object is not used in learning process, however the actual trajectory is also very similar with the desired trajectory. These results leads that the fuzzy feed forward controller can generate the good control inputs and interpolate the control inputs in case of unlearned objects.

Figures 27 and 28 shows the change of fitness values, the number of membership functions, and rules through generations. After 2000 generations, the maximal square error is 1.89×10^{-3}, total square error is 0.361, fitness value is 0.475, the number of membership functions is 79, and rules is 36.

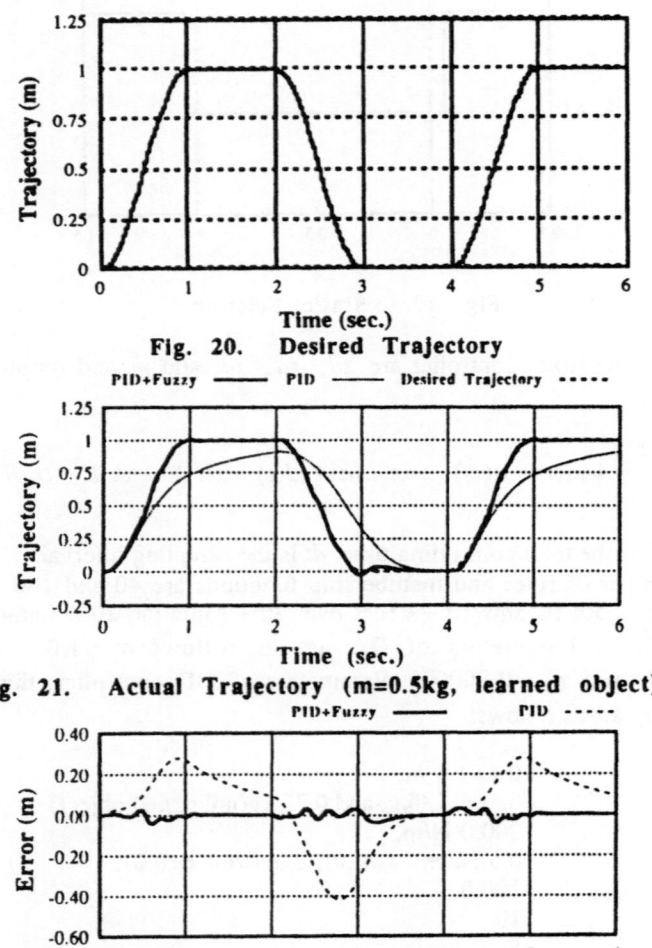

Fig. 20. Desired Trajectory

PID+Fuzzy ——— PID ——— Desired Trajectory ·····

Fig. 21. Actual Trajectory (m=0.5kg, learned object)

PID+Fuzzy ——— PID ·····

Fig. 22. Trajectory Error (m=0.5kg, learned object)

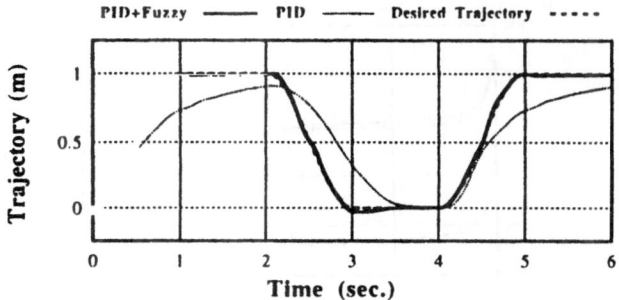

Fig. 23. Actual Trajectory (m=1kg, learned object)

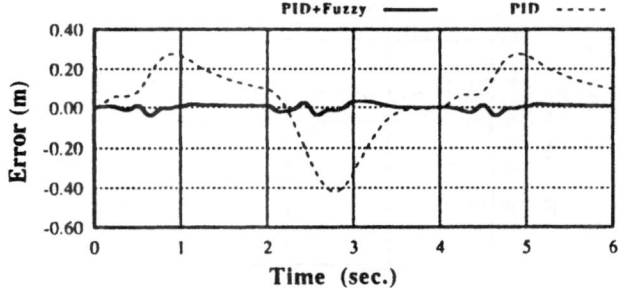

Fig. 24. Trajectory Error (m=1kg, learned object)

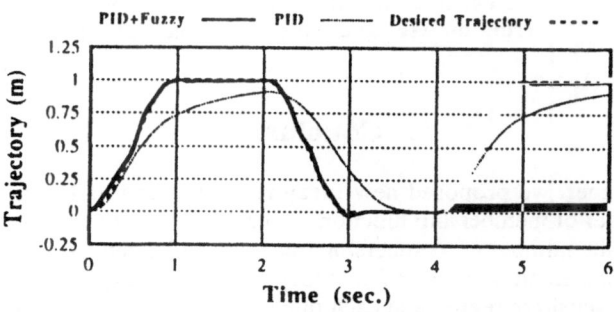

Fig. 25. Actual Trajectory (m=0.75kg, unlearned object)

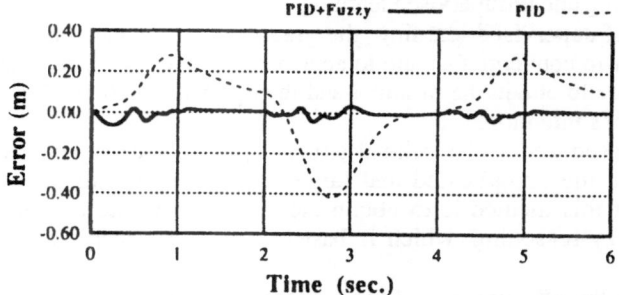

Fig. 26. Trajectory Error (m=0.75kg, unlearned object)

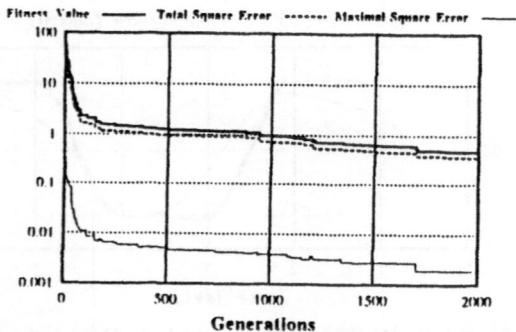

Fig. 27. **Learning Results of Unsupervised Learning (Fitness vs. Generation)**

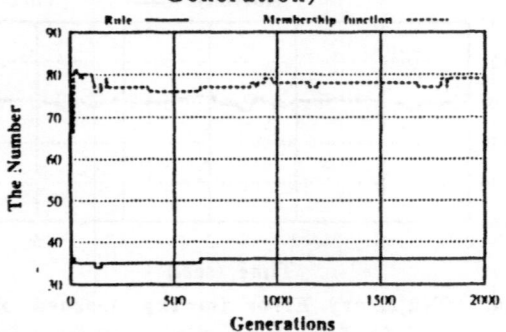

Fig. 28. **Learning Results of Unsupervised Learning (No. of MF and rules vs. Generation)**

5. Conclusions

In this paper, we proposed new fuzzy reasoning and its learning methods to reduce the number of membership functions and rules with high accurate reasoning. In order to reduce the number of membership functions with keeping the accuracy of the system, each membership function of the fuzzy reasoning is expressed by Radial Basis Function with insensitive region and each rule can consist of different combination of membership functions. Genetic Algorithm for learning is classified into two types; supervised learning and unsupervised learning.

In case of supervised learning, the gradient descent method is also used for learning in order to converge fast and to approximate precisely. The characteristics of these methods are to obtain the minimal and the optimal structure of fuzzy reasoning with the operator's intention.

we also proposed a hierarchical structure method for supervised learning in order to prevent the complicated and large fuzzy system with many inputs. The characteristic of this method is to obtain the minimal and the optimal hierarchical structure of fuzzy reasoning, which is based on the relationship among input and output variables.

We show the effectiveness of the proposed fuzzy system. These reasoning can approximate functions precisely with minimum number of membership functions and rules and it can be used for the feed forward control system with good results.

6. References

Fukuda, T., Shimojima, K., Arai, F., Matsuura, H.(1993) Multi-Sensor Integration System based on Fuzzy Inference and Neural Network, *Journal of Information Sciences*, Vol.71, No. 1 and 2, pp. 27-41

Higgins, C., M., Goodman, R., M.(1992) Learning Fuzzy Rule-Based Neural Networks for Function Approximation, *Proc. of IJCNN*, vol.1, pp.251-256

Horikawa, S., Furuhashi, T., Uchikawa, Y.(1992) On Fuzzy Modeling Using Fuzzy Neural Networks with the Back-propagation Algorithm, *IEEE Trans. Neural Networks*, Vol. 3, No. 5, pp. 801-806

Katayama, R., Kajitani, Y., Kuwata, K., Nishida, Y.(1993) Self Generating Radial Basis Function as Neuro-fuzzy Model and its Application to Nonlinear Prediction of Chaotic Time Series, *IEEE Int'l Conf. on Fuzzy Systems 1993*, pp. 407-414

Lee, M.A, Takagi, H.(1993) Integrating Designs Stages of Fuzzy Systems using Genetic Algorithms, *Proc. of Second IEEE International Conference on Fuzzy System*, pp.612-617

Linkens, D.A., Nie, J.(1993) Fuzzified RBF Network-based Learning Control: Structure and Self-construction, *IEEE Int'l Conf. on Neural Networks 1993*, pp. 1016-1021

Nomura, H., Hayashi, I., Wakami, N.(1991) A Self-tuning Methods of Fuzzy Control by Decent Method, *Proc. of 4th IFSA Congress*, Engineering, pp. 155-158

Rumelhart, D.E., McClelland, J.L. and The PDP Research Group(1986) Parallel Distributed Processing, Vol.1, 547; Vol.2, 611, *MIT Press*

Shimojima, K., Fukuda, T., Arai, F., Matsuura, H.(1992) Multi-Sensor Integration System utilizing Fuzzy Inference and Neural Network, *Journal of Robotics and Mechatronics*, Vol.4, No.5, pp.416-421

Shimojima, K., Fukuda, T., Arai, F., Matsuura, H.(1993) Fuzzy Inference Integrated 3-D Measuring System with LED Displacement Sensor and Vision System, *Journal of Intelligent and Fuzzy Systems*, Vol.1 No.1, pp.63-72

Whitley, D., Strakweather, T. and Bogart, C.(1990) Genetic Algorithms and Neural Networks: Optimizing Connection and Connectivity, *Parallel Computing 14*, pp.347-361

Genetic Algorithms for the Development of Fuzzy Controllers for Mobile Robots

Donald Leitch and Penelope Probert

Robotics Research Group, University of Oxford, UK

Abstract. In this paper we present a novel genetic algorithm (GA) for designing fuzzy systems for mobile robot control, in which the meaning of a section of chromosome is determined by surrounding genes, much like a language description. We demonstrate that this leads to much improved convergence speed over conventional GA codings, and discuss the mechanisms that lead to this improvement. The algorithm also uses a simple chromosome reordering operator which uses the algorithm to maximise its own efficiency.

Our results show the performance of our algorithm when applied to designing controllers for the commonly used benchmark inverted pendulum problem, as well as problems in mobile robotics. We discuss the application of this method to robotics generally, and some of the difficulties faced when the algorithm is used for more complicated applications, and how these problems could be approached.

1 Introduction

There can be little doubt that almost all of the concepts we use to reason about our physical world are to a certain extent fuzzy. The paradigm of fuzzy systems seeks to formalise this perception for the purpose of intelligent control. To date, one of the criticisms of fuzzy systems has been the lack of any formal method of deciding how to partition the input space of a system into fuzzy sets, and how to associate these sets with outputs. Many algorithms exist for designing fuzzy systems when an input - output data set exists, such as adaptive vector quantisation [11] and the method proposed by Wang and Mendel [18]. Feedforward backpropagation neural networks can also be thought of as fuzzy classifiers, and supervised learning schemes such as gradient descent evolve means of associating overlapping regions of input space with particular outputs to varying extents, which is in essence what a fuzzy system does.

All such schemes however, as well as unsupervised neural network learning algorithms such as differential competitive learning, require input patterns derived from examples of "good" control. The algorithm presented here uses only an objective function and a simulation to assess the performance of a controller, and a genetic algorithm to improve the performance.

Similar approaches have been presented in the work of Cooper and Vidal [5], Lee and Takagi [16] and Karr [10] amongst others. Our algorithm differs significantly in many ways; our coding scheme is novel and improves the performance of the algorithm significantly, we also include an implicit chromosome reordering operator to reduce non-linearity in the algorithm.

In this section we provide some background information on genetic algorithms and fuzzy logic and discuss their compatibility. Section 2 describes our coding scheme and the chromosomal ordering operator. Section 3 presents the results of the algorithm's application to the inverted pendulum problem, and compares these results to some obtained using more traditional GA / Fuzzy combinations. Section 4 is a discussion of some of the issues involved in applying the algorithm to more complex problems in mobile robotics, and section 5 presents the results of such an application. Finally we conclude with a brief discussion of ongoing research.

1.1 Fuzzy systems

A fuzzy system is basically an expert system which uses fuzzy logic for inferring outputs. Fuzzy logic can be regarded as being a extension of classical logic, the central tenet being that a member of a fuzzy set may have a numerical degree of membership anywhere in the interval $[1, 0]$ as opposed to either 1 or 0 with no intermediate value allowed [19, 15].

The very nature of a fuzzy set makes it impossible to pin down exactly the 'right' membership function, particularly if the concept expressed by the set is multidimensional. In many cases multi-dimensional fuzzy sets are defined by assuming that set membership functions can be expressed in terms of a set of independent one dimensional membership functions. Although this may produce acceptable controllers in many cases, there is no reason to suppose that optimal performance can be achieved with such simplifications. This representation is the equivalent of assuming that a set can be represented as an ideal point in antecedent space, along with a number of shape parameters, and using a L_∞ distance measure in antecedent space when calculating the degree of membership of a particular antecedent vector. In this paper we use the L_2 metric, as this makes it possible to use a much greater variety of set shapes, that are not constrained to be rectangular with their axes parallel to the antecedent space axes.

Fuzzy systems were originally used in controllers because of their power at representing linguistic concepts, such as HIGH, NEAR etc, and their subsequent ability to model the expertise of trained human process controllers. An alternative view is to regard a fuzzy system as a function or control surface approximation. In this sense, the individual sets cannot meaningfully be given linguistic labels, particularly if they are designed using some optimisation algorithm. Linguistic meaning is only applicable when sets are designed with it in mind, and not when sets are described purely numerically as control surface approximators.

1.2 Genetic Algorithms

Genetic algorithms (GAs) are biologically inspired multi parameter search / optimisation algorithms that have proven to be effective at solving a variety of complex problems other algorithms have difficulty solving. In common with all other search algorithms, a GA performs a search of a multidimensional space containing a hypersurface known as the *fitness surface*. A particular parameter set defines a point on a hyperplane on to which the surface is projected, with the height of the surface above the hyperplane reflecting the relative merit of the problem solution represented by the parameter set.

The basis of a GA is that a population of problem solutions is maintained in the form of *chromosomes*, which are strings encoding problem solutions (in this case fuzzy controllers). Strings can be binary or have many possible alternatives (*genes*) at each position. The strings are converted into problem solutions, which are then evaluated according to an objective scoring function. (Similar to a function used to optimise controller gains in LQP control, for example.) Often it is not possible to exhaustively test all aspects of a solution, and noise may be present on the objective function, so the assigned fitness is an estimate of the true fitness of a chromosome. It is important that this is a good estimate, otherwise the *selective pressure* that favours truly high scoring chromosomes can be lost in the noise caused by poor fitness estimates.

Following fitness evaluation, a new population of chromosomes is generated by applying a set of *genetic operators* to the original population. These are basically random copying and altering of individuals from the original population with the probability of copying of any individual from one generation to the next being proportional to its fitness. During the copying process two operations may be performed - a gene may be erroneously copied (*mutation*), or a new individual may be formed by combining segments from two chromosomes by copying one chromosome up to a specific location on the chromosome, then copying a different chromosome (*crossover*). For example, consider the chromosomes shown in figure 1. The two chromosomes at the bottom are the offspring of the upper two, formed by crossover at the line, and mutation of the gene in the box.

GAs are particularly good at finding maxima where the fitness surface is non-linear, highly convoluted, with many local maxima, and dependent on several parameters simultaneously. This non-linear dependency on several parameters which means that overall fitness cannot be expressed as a function of fitness values due to individual genes is known as *epistasis*.

Although in many cases a GA is unlikely to produce an optimum solution to a problem, it can produce a solution that is within a few percent of the optimum in a time orders of magnitude less than a full solution takes using some other algorithm such as exhaustive search. They have found a vast number of applications, ranging from robotic path planning [6] to gas pipeline pumping optimisation [7]. Beasley et al. give many examples of GA applications in [3].

The power of GAs to solve complex problems can be viewed in terms of the *schema* theorem first proposed by Holland [9]. A schema is a short section of a chromosome, which need not necessarily be contiguous. Holland's theorem

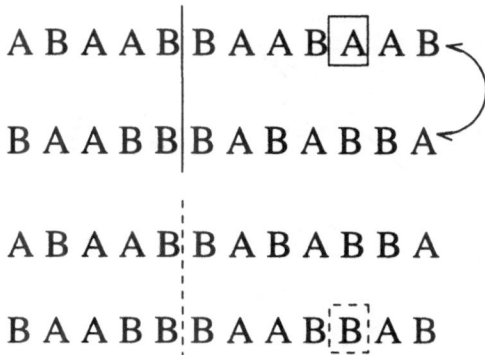

Fig. 1. Genetic operators

basically shows that schema which when part of a chromosome lead to a good score, are likely to increase in frequency throughout the population, thus leading to a gradual improvement in fitness as time passes under certain circumstances such as very large populations.

Although the schema theorem is useful, it ignores the effects of epistasis and makes many other simplifications and assumptions, and other suggestions have been put forward for explaining the success of genetic algorithms, such as Altenberg's work on the evolution of evolvability [1], and Goldberg's building block hypothesis [7]. Goldberg's book is also a good general introduction to GAs.

Altenberg has produced theoretical analyses of genetic programming (GP), which are particularly useful when analysing the coding scheme described in this paper, as there are many similarities between GAs using the scheme and the paradigm of GP, where tree like logical and arithmetical structures are adapted to produce compound functions and programs. Although the full details are too lengthy to be included here, the essence of the theory in [1] is that a property called *evolvability* arises, which is the ability of particular sections of chromosome not only to contribute towards overall fitness, but also to be good at combining with other sections of chromosome to produce blocks of greater fitness, thus not only is GP good at finding blocks of code which combine well by chance, mechanisms exist by which versatile and highly fit blocks of code arise as a property of the algorithm.

1.3 GAs for the evolution of fuzzy systems

If we imagine a fuzzy rule to represent a patch on a control surface (figure 2), it is easy to visualise why GAs are particularly suited to the design of fuzzy systems.

The output for a given input will generally only be dependent on a single rule. Thus if the parameters on the chromosome which code a particular rule are changed, the input / output function defined by the chromosome's corresponding rule base will only be affected in a small area. If the input falls exclusively

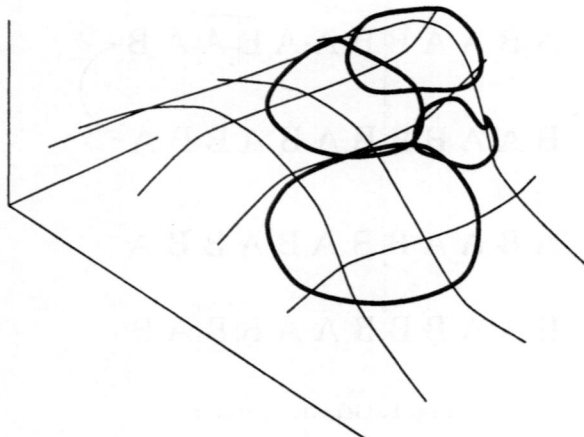

Fig. 2. Fuzzy rules as patches

within a single set, then the effect in that region of changing the chromosome is not affected by any other parameters on the chromosome. Only when an input falls in a region in which two or more sets overlap is there any interdependency between chromosomes, in that the effect on the output of changing one parameter is influenced by the value of another parameter. This is epistasis, and by the very nature of fuzzy systems, is limited. As observed by Davidor [6], GAs are particularly well suited to problems with this kind of limited epistasis.

2   Genetic coding of fuzzy rule bases

The way in which a problem solution, in this case a fuzzy rule base, is represented in the chromosome, is of great importance for the success of the GA [7]. Many schemes suggest themselves for representing a fuzzy rule base using a string of numbers, depending on the amount of freedom we wish to give the GA to alter the rule base. For example, the simplest possible application of a GA to learning a rule base is one in which the input and output sets are predefined by the user, and the GA is simply used to learn associations between them. In this case, the rule base could be coded as a string of integers. This method was tried with some success by Karr [10], but it relies on human expertise to partition input and output spaces, and so slightly defeats the purpose of using the GA in the first place. A more complex method uses a fixed number of parameters to describe set positions and shapes, with the number of rules in a rule base determined in advance. This coding scheme has been widely applied, but still relies on a initial guess as to the number of input space partitions necessary to achieve a good problem solution. Ideally the number of rules as well as their shape could be decided by the GA. For this to occur, however, chromosomal length has to be variable and crossover has to be able to occur at different sites on parent chromosomes.

This can cause problems, as it becomes possible to generate "impossible" chromosomes that cannot be converted into rule bases, unless measures are taken to ensure crossover occurs at legal sites, or that impossible chromosomes are either weeded out or penalised by the objective function. It may also be necessary to constrain the two crossover sites so that chromosomal length does not vary too much. Despite the increase in complexity over other coding methods and the need for support structures to ensure chromosomal viability, methods using variable length codings have been used with success, for example Cooper and Vidal [5] and Lee and Takagi [16]. Examples of more complex coding schemes that do not rely on a string of real numbers as parameters specifying rules can be found in Nomura et. al. [14] who use an interesting binary method where the fixed length chromosome is a direct representation of the input universe of discourse, and features on the chromosome translate directly to features on the universe of discourse, eg set centres.

2.1 Context Dependent Coding

The coding scheme adopted in this paper uses language like strings to encode a rule base, which are interpreted by a parser. The parser has a limited degree of intelligence, and knows what sort of structure the rule base takes. Particular codons in the chromosomal string indicate the context in which sections of string should be interpreted. An example of a general CDC scheme is described, followed by a specific example of CDC applied to the coding of fuzzy rule bases.

General Scheme A general CDC scheme might use four codons, two of which are used to represent instructions, and two of which are used to encode numbers. This way it is possible to represent any instruction by indexing it according to its number which is represented in binary, or code any integer in binary. It is not necessary to use two different codons to represent instructions or numbers; two is just the minimum number that can be used for a general scheme.

The parser operates by scanning a chromosome from left to right, assigning meaning to each codon or group of codons depending on the context in which it finds them. Consider, for example, a system which can be described using four instructions, plus numbers. Each instruction has a number of real values associated with it, and may also have further instructions associated with it. Using a four letter alphabet of codons, assume that A and B are used for instruction coding (*instruction codons*), and 1 and 0 are used for number coding (*numerical codons*). As there are four instructions each one will be coded by a sequence of codons rather than by a single codon. A typical chromosome will be a random sequence of these four codons, of indeterminate length, for example

A1BA0100B10AAB011B00B1A

Initially the parser starts from the left, looking for an instruction, which will be coded as AA, AB, BA or BB, as there are four instructions in the language. The parser ignores the initial two codons, and finds the sequence BA, which is assumed to have some meaning to it in this case. It then may need to look for a numerical sequence, so it will read the sequence of numbers following the BA, terminating it when it finds another instruction. In this case the sequence is 0100B10 before the next instruction , AA, is encountered. The B is ignored and the binary sequence 010010 is converted in to a number by whatever scheme is being used. Following the AA instruction is a B, which is also ignored. The next sequence may be decoded as a number, or ignored, depending on the meaning of instruction AA. The process then continues until the end of the chromosome is reached, at which point any unfinished instructions are either discarded or terminated with a standard sequence. By careful design of the instruction set, any chromosome can be decoded no matter how long, or what sequences of codons appear within it. The simple scheme described above can be varied in numerous ways. For example once an instruction codon has been encountered when looking for an instruction, any numerical codons can be ignored; in the above example the first instruction would be AB instead under this scheme, which is useful if the instruction set is large, as long sequences of letter codons are unusual. Another modification may be that any letter symbol terminates a numerical sequence, this has the effect of increasing the number of codons which are ignored (assigned no meaning) by the parser.

Although the scheme may not be very elegant, it is very flexible and robust, and has several advantages over conventional codings which are enumerated later in this paper.

Coding of a rule base Consider a four input, two output system using sets described by a centre in input space and a radius. Due to the simplicity of the system, it can be described by a string utilising a three letter alphabet, two letters of which (0 and 1) are used to code real valued numbers or binary switches (*binary codons*), and one of which (C) is used as a *context switch* (instruction codon), to indicate to the parser the start and end of particular sections of code. For example, take the section of code shown in figure 3.

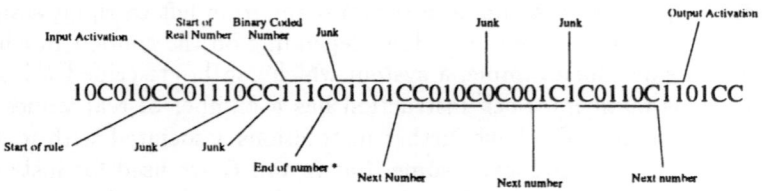

Fig. 3. Section of CDC Chromosome

As the parser scans the chromosome from left to right, initially it is looking for CC to indicate the start of a rule. The parser then knows that for a four input system, it must read the next four binary codons to indicate if each of the four inputs is used in this rule (1) or not (0). In this example, section following the two adjacent context switches is 01110, the first four codons of which indicate that this rule applies to inputs 2, 3 and 4. Had the parser encountered any context switches is the middle of this section, they would have been ignored as they have no meaning in this particular context. The parser then knows it has to look for three coded real values, to indicate the position in input space of the centre of this rule's antecedent set. The start and end of real numbers are marked by single Cs, so the parser ignores the 0 immediately following the four binary codons indicating input activation, and the next C indicates the start of a real number. The adjacent C is ignored, as although a C is used to indicate the end of a number, no binary codons have been found yet. The following string of binary codons, 111, codes the number 1.0000, according to the following scheme. For a string of length n codons (this is obviously variable, as the terminating C is not fixed), the corresponding real number is $\frac{k}{2^n - 1}$, where k is the value of the binary number. Due to the variable length, the precision is variable. All numbers are scaled to be in the interval [0,1].

Although the use of Gray coding in binary representations of real numbers is widespread, use of such a scheme is only of benefit when a position dependent coding is used. This is because a GA using a position dependent coding is optimising a fixed length string of parameters, which are always in the same place on the string. In a CDC, the parameters determining fitness are distributed in various different positions on the string and so when two parameters happen to be mixed by crossover, it is almost certain that the new parameter formed will be unrelated in the effect it has on fitness to the original two parameters.

The five binary codons following the terminating C are ignored, as the parser looks for another C to start the next real number, the position in input space of the centre of the antecedent set with respect to the third input to the system (second input to the rule). The number is decoded in the same way as before, and has the value of $\frac{2}{7}$. The final set centre position is $\frac{1}{7}$. The radius of the conical set in input space is decoded as $\frac{6}{15}$. Obviously it would be easy to modify the parser to allow for more set shape parameters, and even different geometries.

The parser has now got all the information describing the input set for this rule, it now looks for information concerning the output part of the rule. First, it looks for another binary sequence to indicate which outputs this rule affects. The next sequence after the section coding the radius is 1101. The first two 1s indicate that this rule affects both outputs. The next two codons are junk, then the next C indicates the start of the real number indicating output 1 if this rule is fired.

It can be seen that a considerable portion of the chromosome does not code any useful information. This may seem inefficient, but these *junk* genes play a useful part in absorbing disruption caused by genetic operators. Consider, for example the effect of a mutation at the codon marked *, say a change to a 1. This results in the first set centre position being decoded as $\frac{461}{512}$, as the following section of junk becomes part of the number. Following that, however, the decoding remains the same. Mutations in other positions have differing effects, but generally the damage is limited. The same applies to crossover; the meaning of the chromosome is disrupted in the region of the break, but it is unusual for more than a single rule to be affected. If some algorithm were implemented to remove the junk every time a chromosome was parsed, genetic operators would become highly disruptive and would be likely to change the meaning of an entire chromosome.

Other advantages of CDC are that the variable length and independent crossover sites on each chromosome allow for much greater genetic variability, as two sections of chromosomes that occupy identical sites on different chromosomes can find themselves on the same chromosome as a result of crossover, which cannot occur for a fixed length position dependent coding.

Furthermore the presence of junk has an effect on efficiency noted by Levenick [13], who introduces *introns* (crossover targets) in to chromosomes and observes a considerable increase in efficiency. Angeline [2] also notes that when using GP redundant sections of code often arise, and deleting these seems to have a negative affect on the performance of the algorithm. It is possible that the presence of junk allows the chromosome to adapt to become robust to disruption by crossover, so that the second order effects predicted by Altenberg [1], namely that not only do good chromosomes arise, but structures within chromosomes that enable them to be more adaptable in the presence of crossover, arise as well.

The greater genetic diversity and unpredictable disruption caused by the genetic operators also helps to reduce premature convergence, as they make it extremely unlikely that all the individuals in a generation become very similar, and even if they are all identical novel chromosomes are still generated from the current population.

The coding is also very flexible; systems far more complicated than rule bases such as recursive networks can easily be described with modifications to the parser, and a general scheme capable of describing any system or program can be realised using four codons, with two used for instruction coding and two for numerical coding.

2.2 Chromosomal Fitness

As mentioned in section 1.3, the interaction between sections of chromosomes representing individual rules is limited as the interaction between rules is limited. In a randomly generated rule base, however, rules that have antecedent set centres close together in input space and are likely to interact are not necessarily close to each other on the chromosome. The result of this is that the crossover operator can be very disruptive, with offspring strings producing behaviour that differs greatly from both of their parents due to large numbers of interactions between the sections of chromosome that came from each parent. This can be advantageous, as it leads to the occasional development of novel behaviours, but it is detrimental to the incremental improvement in behaviour that is due to offspring having the characteristics of already successful parents.

This can be combatted by using chromosomal ordering algorithms, for example [10] describes a heuristic method of aligning strings before crossover to maximise the exchange of similar material, and Hoffmann uses an ordering operator in [8] to reduce the disruptive effect of crossover.

157

The parent-offspring behavioural continuity is more likely to be maintained if sections of chromosome that represent sets that are likely to interact are as close as possible on the chromosome. Crossover is unlikely to disrupt such clusters, and also unlikely to bring together sections of chromosome that interact and produce unexpected behaviour. What is therefore desired is some ordering operator that manipulates sections of chromosome. If the centres of the rules are thought of as being points in a space with a dimension equal to the number of inputs (figure 4), a chromosome, in ordering these points, describes a path through this space. In simple terms, what we wish our ordering operator to do is to minimise the length of this path.

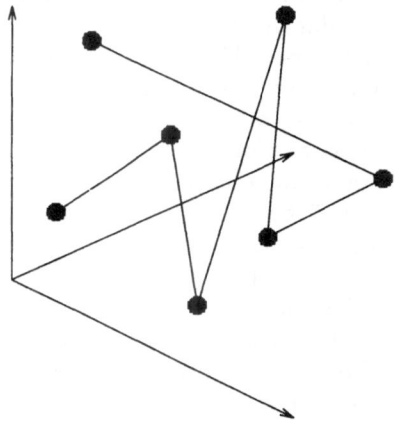

Fig. 4. Chromosomal path through input space

This is in fact the well known travelling salesman problem (TSP), a complex NP-complete problem that has no analytical solution for a large number of points. A successful applications of genetic algorithms has been in producing solutions to the TSP [3]. Using a CDC, however, it is very difficult to change the order of a chromosome without badly disrupting it, so explicit ordering is not feasible. Instead, it is possible to do it implicitly using the GA itself. The method is simple - an approximation of the degree of epistasis is made by taking the sum of the distances between the centres of antecedent sets that are adjacent on the chromosome, weighted by the radius of the sets. The reciprocal of this measure, called the *chromosomal fitness*, is included in the objective function, so that one of the goals of the GA is to minimise this measure.

If N set centre positions in input space are given by the vectors $S_1, S_2, \ldots, S_n$, their radii $r_1, r_2, \ldots, r_n$ then the chromosomal fitness is given by

$$\frac{1}{n} \sum_{i=1}^{n-1} u(0)(r_i + r_{i+1} - ||S_i - S_{i+1}||)$$

Where $u(0)$ is the unit step function, included so that two sets do not interact by a negative amount if they are a long way apart in input space.

More complex measures can be used, for example calculating some weighted sum of the distance between sets that are close on the chromosome but not adjacent, with the weight being inversely proportional to the distance between the sets on the chromosome, or by calculating chromosomal fitness to account for the volume of input space occupied by set overlaps instead of using the linear method shown above. It was found experimentally that for the applications discussed in this paper that more complex measures than the one shown did not provide a significant advantage.

It was found that the best way to combine chromosomal fitness and raw fitness in the objective function was to multiply them together following scaling to ensure that they had similar geometric ranges (i.e. the ratio initial value to final value was about the same). This implicitly rewards the algorithm for improving the attribute that is worst, as a unit increase in the smaller quantity produces the greatest increase in overall fitness. Although other methods of combining multiple objectives are often used, for example Pareto optimality [7] simple multiplication was found to be effective.

It also avoids the problem of the algorithm becoming totally concentrated on improving one attribute, which sometimes happens if they are summed. This is true of any case where a GA has to improve a number of attributes in parallel.

Using this technique avoids the need for time-consuming explicit reordering algorithms and enhances the GAs natural power of self optimisation.

The results of using a chromosomal fitness measure are shown in section 3.

Although the use of chromosomal fitness is a disadvantage initially as the GA will be biased towards some strings purely on the basis of their structure and not their fitness as regards the behaviour of the corresponding rule base, this is more than compensated for later by the increased efficiency of the algorithm due to the order of the strings.

3 Algorithm applied to Cart-pole balancing

In this section we present the results of applying the algorithm to the inverted pendulum stabilisation problem, often used as a benchmark for controllers. We present comparisons with standard GA methods used for generating fuzzy rule bases, and briefly with another rule based method [17].

The cart-pole problem (see figure 5) is interesting because two variables have to be controlled using a single control input. There are four system states, and the dynamics of the system are governed by the following equations [5]

$$\ddot{x} = \frac{F - \mu_c sign(\dot{x}) + \tilde{F}}{M + \tilde{m}}$$

$$\ddot{\theta} = -\frac{3}{4l}(\ddot{x}\cos\theta + g\sin\theta + \frac{\mu_p\dot{\theta}}{ml})$$

where

$$\tilde{F} = ml\dot{\theta}^2 \sin\theta + \frac{3}{4}m\cos\theta\left(\frac{\mu_p\dot{\theta}^2}{ml} + g\sin\theta\right)$$

$$\tilde{m} = m\left(1 - \frac{3}{4}\cos^2\theta\right)$$

where the cart mass $M = 1.0$ kg, pole mass $m = 0.1$kg, pole half-length $l = 0.5$m, friction of cart on track $\mu_c = 0.0005$N, friction at hinge between cart and pole $\mu_p = 0.000002$kg m, cart position is x and pole deviation from vertical is θ. For comparison, the parameter values are taken from [5], but the simulation used here differed in that it included noise and had a lower sampling rate (20 Hz as opposed to 100 Hz).

There are four control inputs - θ, $\dot{\theta}$, x and $\dot{x}$. The control output is the lateral force applied to the cart.

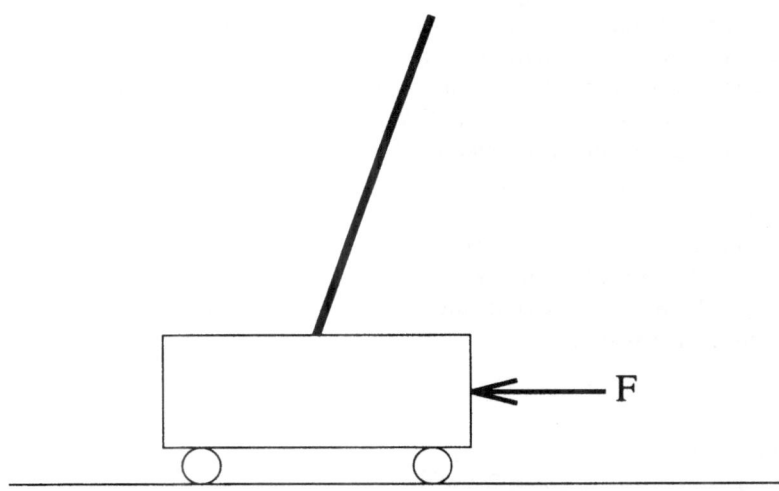

Fig. 5. Cart-pole system

The system was simulated on a computer, using an Euler scheme to calculate derivatives. Random noise was added to simulated sensor inputs to the controller, and to the actuator outputs. This was done in order to more closely model the real world, and also so that the controller was robust enough to cope with the inevitable differences that would result from transferring it from a simulated environment to a real one. The presence of noise actually enhanced learning as well as leading to more robust controllers [12].

The simulation had an update rate of 20 Hz, as did the controller. The scoring function used to assess the fitness of solutions over a trial was:

$$S = \sum_{test}(1 - x^2)$$

.expressed as a percentage. If the controller failed (defined to be $\bmod x > 1.0$ or $\bmod \theta > 0.26$) the trial was stopped, so that a controller which failed after only 10 % of the maximum allowed time could only score a maximum of 10 %. Each controller was tested for up to 40 simulated seconds, from 12 different starting states. The best controller in each generation was also tested for 10 minutes from a random starting position, to test robustness and generalisation capability. A controller capable of stabilising the pendulum whilst keeping the base within the specified tolerance for the long test was usually evolved within the first 10 generations. The population was 30.

3.1 Chromosomal fitness

Figure 6 shows a comparison of GA convergence rates with (curve A) and without (B) a chromosomal fitness term included. The curves shown are of raw fitness (chromosomal fitness was only used to determine the probability of copying to the next generation), and are obtained by averaging the best score in each generation over six independent runs, as a single GA run is very noisy.

Three curves are shown; the two upper curves are of GAs using chromosomal fitness, and the lower curve is of a GA not using chromosomal fitness. Two versions of the GA using chromosomal fitness were included to illustrate that the difference between use and non-use of chromosomal fitness is significant and not random.

The population used was 30, the mutation rate was 0.005 (the probability that a single codon would be subject to mutation, not the probability of changing a codon) and the probability that any pair of chromosomes selected for copying were crossed over was 0.6.

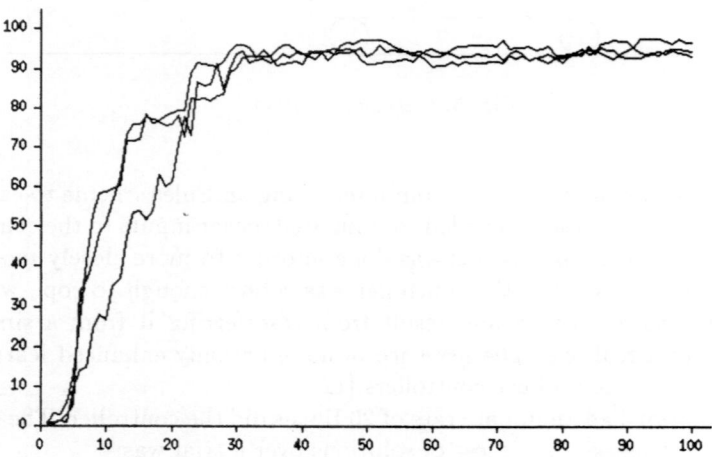

Fig. 6. Comparison of convergence rate with and without chromosomal fitness

3.2   GA parameter settings

The parameter setting of the GA were varied to examine the effect that this would have on convergence. Figure 7 shows the convergence curves obtained using mutation only, with the mutation rate set to 0.05, 0.005 and 0.0005. The best convergence is obtained with the rate set to 0.005, and the very noisy curve is with the highest mutation rate. It can be seen that mutation is actually a fairly efficient search operator.

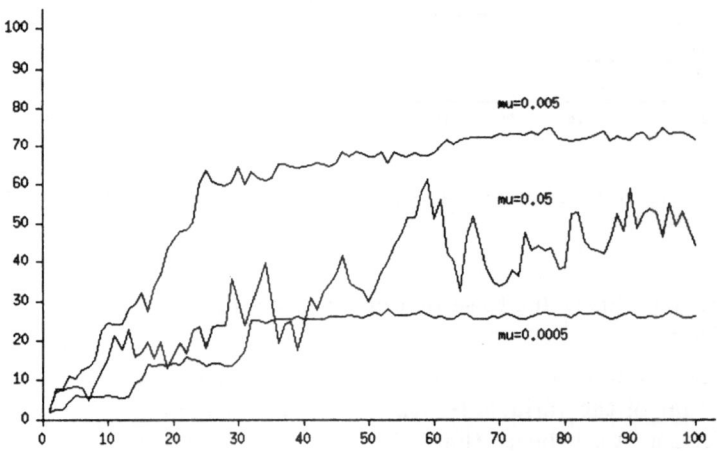

Fig. 7. Convergence of GA using various mutation rates

The effect of varying crossover is shown in figure 8. The two curves shown are produced by GAs with crossover rates of 1.0 (upper curve) and 0.2.

3.3   Comparison with standard coding

Figure 9 shows the convergence of standard GAs when applied to the same problem. The curve that quickly converges is a fixed length, real coded chromosome with the number of rules fixed at 20 (with CDC the number of rules was usually in the range 10 - 30), and the plot shown is the mean of six runs, again using the best score of each generation. The other curve is obtained using a similar representation, but with a variable length chromosome, and crossover sites constrained so that viable chromosomes are always produced.

It can be seen in figure 9 that although allowing the length of the chromosome to vary and picking independent crossover sites on each chromosome improves performance when compared to a standard GA, the GA using CDC significantly outperforms the variable length GA, indicating that some other mechanism than increased variability due to chromosomal mixing is involved in the improvement.

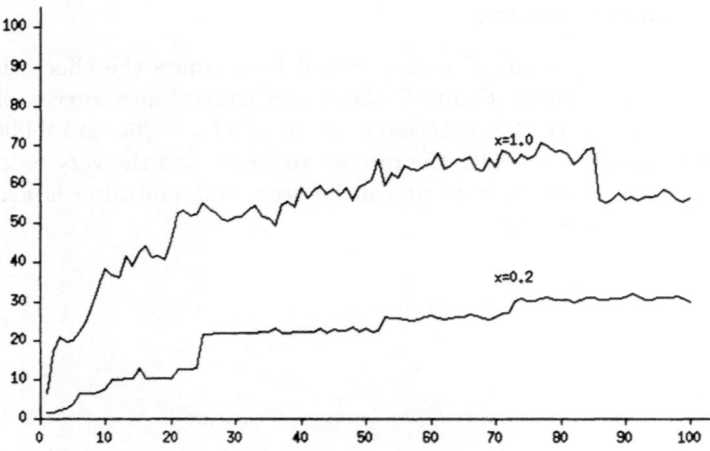

Fig. 8. Convergence of GA using various crossover rates

It is thought that this is due to second order effects that become possible in the presence of junk.

It is interesting to note that the convergence of the fixed length GA is initially faster than that of the variable length GA, although the latter eventually surpasses the former. It is believed that this is caused by the latter's far larger search space, and subsequently the initial period is spend building up small blocks of good rules which then start combining to produce the increase in convergence speed seen later on.

Cooper and Vidal describe a GA for evolving fuzzy inverted pendulum stabilisers in [5]. As a performance measure, they use the ability of the controller to stabilise the pendulum within certain limits for a long (hours of simulated time) trial. Their algorithm could achieve this in 10000 function evaluations, our algorithm frequently produces controllers capable of long term stabilisation within 1000 function evaluations. A controller is deemed to be capable of long term stabilisation if it can balance the pendulum (within the limits mentioned earlier) for a period of 24 hours of simulated time.

Varšek et. al. describe a rule based controller attributed to Makarovič [17], which in our trial had a fitness of 95% when the parameters were tuned by hand. Not only do many of the controllers produced using the algorithm described here outperform this, Makarovič's controller, and the controllers similar to it that use GAs for parameter determination in [17], imposed rigid structures on the input space, either using fixed partitioning, or decision trees to rank state variables. This type of structure would not be suitable for problems with a high dimensionality, or problems with a large degree of asymmetry, such as the mobile robot controllers described in the next section.

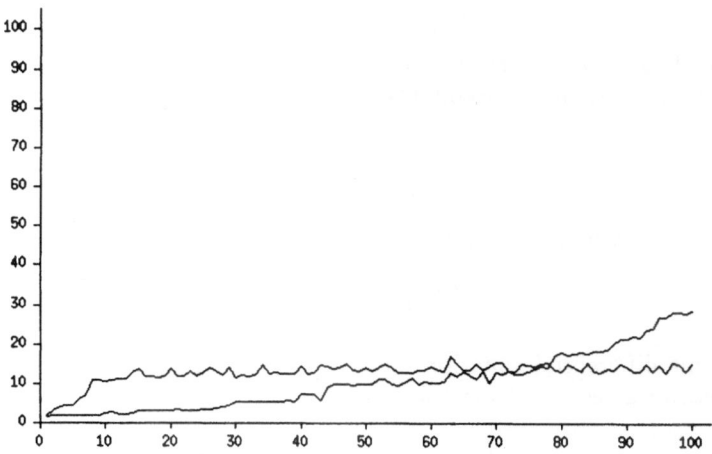

Fig. 9. Convergence rate of standard GAs

4 Problems in Robotics

Mobile robots provide a far greater challenge than simple 'toy' control problems. In the inverted pendulum, the system state can be measured directly and only a single control output is inferred. The task of balancing is also very easy to specify. A mobile robot has an array of sensors, which lead to a redundant, non-linear mapping from sensor inputs to states. There are normally at least two actuator outputs to be inferred from the input data, and the relationship between current state and next state is often highly configuration dependent and non-linear. Unpredictable environmental changes also affect the robot.

It is also far harder to classify behaviour quantitatively for a mobile robot which may need to execute complex manoeuvres than for a simple controller whose behaviour consists of tracking a set point.

The objective function used by the GA is the only method of judging the fitness of a behaviour. Good choice of objective function is therefore of paramount importance to the success of the algorithm. In this section we briefly analyse the application of a simple objective function, similar to the one used in the cart-pole problem, to robotics.

Let the state of a robot with respect to its environment at time t be $\mathbf{x}_t$. Define also the sensor reading vector to be S_t, E the state of the environment, which is entirely external to the robot, but with respect to which a task is defined, and the actuator output A_t.

These quantities are related through the equations:

$$S_t = f_1(\mathbf{x}_t, E)$$

$$A_t = f_2(S_t)$$

$$\mathbf{x}_{t+1} = f_3(A_t)$$

which are likely to be very non-linear and redundant. f_2 is the controller function; the only way of modifying fitness is by modifying this function. More concisely:

$$\mathbf{x}_{t+1} = f_1 f_2 f_3(\mathbf{x}_t, E)$$

this can be written as the timestep $\Delta t \to 0$ as a differential equation:

$$\frac{d\mathbf{x}}{dt} = g(\mathbf{x}, E)$$

Solving this differential equation for a given set of boundary conditions $\mathbf{x}_0$ (the initial state) gives us a trajectory through state space:

$$\int_{trial} \frac{d\mathbf{x}}{g(\mathbf{x}, E)} = \int dt$$

let:

$$\mathbf{x}_t = h(t, E)$$

This trajectory is of course affected by noise in sensor readings and actuations, and changes in the environment, but there is nothing that can be done about this other than ensuring that the controller is robust enough to deal with uncertainty.

The trajectory that results is also highly dependent on the boundary conditions of the equation. From this trajectory, the fitness of the controller must be calculated. For control surface approximation the ideal measure of fitness would be some integral of deviation from an ideal control surface. As we have no idea what the ideal control surface looks like, this is not possible and we have to infer error from indirect means, such as defining an ideal trajectory through state space and measuring the sum of the deviation from this - thus giving us a fitness measure for a single trial:

$$fitness = \int_{trial} |B(t) - h(t, E)| dt$$

Where $B(t)$ is the ideal trajectory through state space for given boundary conditions, or in simple cases a set point. This is obviously a very indirect method of measuring the fitness of a surface, and can lead to many complications. As noise and small changes in boundary conditions may lead to large divergences in state space locii, very different scores may be obtained for similar controllers which can make the problem GA deceptive. One way round this is to conduct a large number of short trials, avoiding divergence. As all controllers are run on a computer, they are in fact using discretised versions of the equation above. Using a large timestep allows large jumps to be made over the control surface, avoiding discontinuities which also cause divergence. The problem which presents the greatest difficulty, however, is the choice of $B(t)$, the desired behaviour.

4.1 Behaviour modification using GAs

Simply specifying $B(t)$ as a trajectory through state space is not good enough, as a perfectly good controller may well produce an identical trajectory at a different rate to the one specified. Some sort of dynamic programming algorithm may be able to match actual and desired behaviour independently of time, or a neural network could be trained to recognise 'good' behaviour, but that defeats the object of using a GA; that no training data need be supplied. Furthermore, the desired trajectory through state space is likely to be dependent on the initial conditions, and will need to adapt to account for changing environments. Apart from very simple cases, for example tracking, it is highly improbable that a designer could design an objective function that specified all aspects of a robot's behaviour under a range of conditions.

One possibility is to use proximity to final desired state as an objective function, but for most tasks this is far too simple as it will not reward partial solutions if they happen to fail in an area of state space distant from the intended final state.

The problem of objective function specification can be viewed as trying to specify a particular behaviour that is desirable, where a behaviour is simply a locus or set of locii in state space.

A fuzzy rule can be thought of as causing a very simple behaviour in an AGV, or defining a short locus through state space, usually only in a specific area of state space. The emergent behaviour caused by all the rules acting together is the locus that the robot follows.

The traditional approach to designing robot controllers to achieve a certain behaviour is to decompose the high level behaviour required in to lower level, simpler behaviours. It is hard to see how some equation defining a set of locii could be decomposed in to simpler equations, but behavioural decomposition can easily be done intuitively. For example, a manufacturing operation may be very hard to define using a single equation describing a locus or locii in the robot's state space, but can easily be described using terms such as "pick up" and "insert", which are lower level behaviours than "assemble object". In a sense low level behaviours are like sub goals that must be achieved in order to attain some global goal.

To achieve a high level behaviour by manipulating rule bases with a GA it is necessary first to develop low level competences which can then be further modified and enhanced to produce an emergent high level behaviour.

This suggests an evolved version of a Brooks type architecture [4] with a GA learning various behaviours at the lowest level by manipulating individual rules, and also at a higher level by manipulating combinations of lower-level controllers.

A GA could build a high-level behaviour on the basis of successfully evolved simpler controllers, as fuzzy rule bases can be combined in the same way as individual rules. A GA that can successfully manipulate single rules will be likely to be able to manipulate pre-evolved groups of rules which already implement simple behaviours. For this to happen, a two level evolution procedure is necessary.

Initially the GA learns easy to define low level behaviours, and once a reasonable degree of proficiency has been attained, the low level rule bases can be manipulated by the GA to form a controller capable of performing a higher level task. A simple objective function is sufficient as the controllers would already have a degree of competence at the behaviour desired.

This kind of hierarchical design assumes operator influence to choose intermediate level behaviours, but without this the GA would be unable to solve the complete problem, or to bridge the 'complexity gap' between the high-level behaviour desired and the low level behaviours it has available to manipulate. What is needed is some kind of operator supplied 'stepping stone' in the form of intermediate level behaviours. It is likely this is how complex behaviours arose in biology; humans and higher vertebrates didn't appear overnight, rather the complex behaviours exhibited by such creatures are based on simpler ones inherited from their ancestors, but subsequently combined and modified to produce an emergent, higher level behaviour.

One of the applications in the next section demonstrates how this multi level controller design may be achieved in practise.

5   Applications in robotics

In this section we present some results obtained using the algorithms and techniques discussed above. Two behaviours are learned using the genetic algorithm; a simple controller which moves along a corridor using simulated sonar readings, and a controller which executes a multi point turn. The robot moves in a simulated corridor environment, using eight sonar, two each on the front and rear and two on each side, to judge distance from the walls. The sonar are simulated bore sight, including noise and specularity. The robot has two independent driving wheels, and the control outputs of the system are signals controlling the angular position of the axles. This is only a kinematic simulation of a dynamic system. However, the dynamics were taken in to account by using a heuristic to calculate if the vehicle was likely to skid from the difference between two sets of sequential control signals, and reducing the fitness of any controller which caused excessive skidding.

Although simulations have to be used with caution when developing controllers as inevitably many real world effects are unmodelled, this particular simulation has been used in the past to develop controllers which have subsequently been successfully implemented on a real robot.

5.1   Corridor tracker

This is a very simple application of the algorithm, as only a single level of behaviour is learned, and it is easy to specify in a number of ways.

Following trials, the best measure of fitness was found to be distance moved parallel to the axis of the corridor before collision, rather than some summed measure of perpendicular distance from the centre of the corridor. The length of a trial is limited as controllers quickly arise which are capable of avoiding the walls indefinitely. This method of assessing fitness avoids the integral of section 3 by assessing behaviour according to final position in state space. A consequence of measuring fitness entirely by distance travelled is that the robot does not try to stay in the centre of the corridor, and an off-centre path scores as well as a centred one as long as this does not lead to excessive collisions, as can be seen in figure 10.

Paths produced by a typical controller are shown in figure 10. The controller came from the 40th generation of a GA run with 30 individuals, requiring about 10 minutes of CPU time to run on a sun4. Although the controller had no knowledge of previous inputs, large changes in control signals can be avoided by keeping the control surface fairly flat, and the algorithm produced controllers which did not skid at all in a few generations.

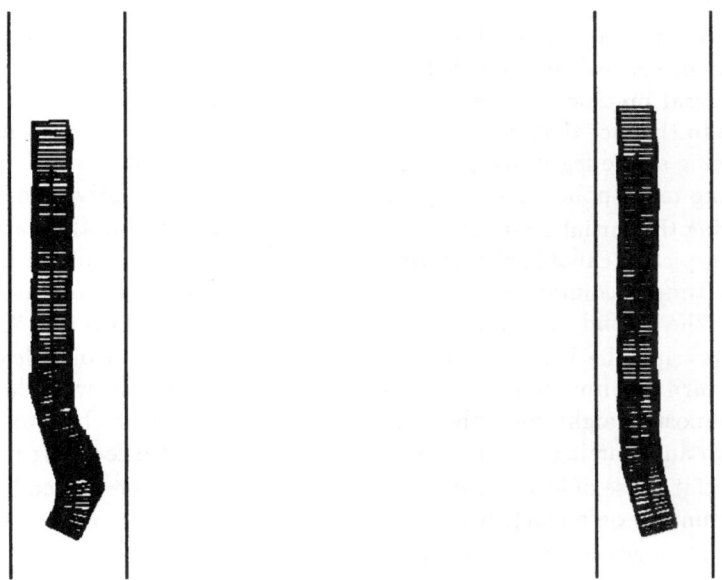

Fig. 10. Paths produced by evolved corridor tracker

5.2   Multi point turn

Performing a multi point turn is a complex manoeuvre, not least due to the fact that the control surface must have discontinuities. It was found that the GA was unable to learn a multi point turn starting from scratch using only total distance turned over a number of time steps as an objective function, so the behavioural approach to learning was used. The GA was run in the normal manner, with the objective function changing during the run to switch from learning simple local behaviours to learning the more complicated global behaviour. The initial phase of learning consisted of testing each controller from a number of starting points in state space, and allowing the controller to evaluate a single step (which could be quite a long move). The score of the controller was then based on how much the bearing of the vehicle had changed, minus a heavy penalty for colliding with a wall. The penalty for skidding was suspended, as the controller could be thought of as planning arcs rather than being a reactive motion controller. The controller effectively learned many simple behaviours in parallel, each of which would obtain the maximum change in bearing from a particular position. When combined these behaviours cause a five point turn to be performed.

A second phase of learning was then done, in which each controller was tested for a longer trial, and fitness was assessed purely on the basis of final state after a number of steps. The second phase of learning was started when no further progress was being made in the first stage, that is, the algorithm appeared to have converged.

When the second phase objective function was used from scratch, it was found that the controllers produced would learn how to perform a single leg of a turn with great precision, but never got any further than this. It is thought that this is due to the fact that the algorithm could initially get large gains in fitness by optimising single leg moves, but the controllers so produced were incapable of producing multi point turns; a good example of a GA deceptive function.

Following the initial phase of learning, the best controller in the final generation was typically capable of performing a five point turn 50 % of the time, the rest of the time becoming "stuck" half way round rather than colliding with a wall. After the second phase of learning, the success rate was over 85 %.

For this trial, the region of state space for which the controller learned to perform a turn was limited to the central regions of the corridor, with the vehicle pointing almost straight down the axis. For a controller to learn how to perform a turn from any starting position would require an extended learning period.

The initial phase of learning needed to produce the controllers used here took about 30 minutes on a sun4, with the second phase taking about the same.

Figure 11 shows some typical turns.

The convergence rate of the GA can be seen in figure 12, which is the mean logarithm of average fitness of four GA runs. After 100 generations, the objective function was changed to assess the performance of each controller as a multi point turner rather than to assess them as capable of performing simple low level behaviours. The curve shows the fitness as assessed, with a scaling factor used in the second phase of learning to maintain continuity of fitness (this is purely to improve the appearance of the graph). It can be seen that the learning rate increases when the objective function is changed, having converged in the first phase of learning.

The two stage learning process is based on the assumption that a GA, like any system that learns, can learn a complex process by breaking it down and learning the components in parallel. Once this has been achieved, improving the whole process can be undertaken.

Fig. 11. Multi point turns performed by evolved controller

Obviously the learning process is not limited to two stages, there may be several levels of behavioural decomposition possible, or necessary. At present a human operator must decide on the way the problem is decomposed. However, with large class of problems this is not difficult. Broadly speaking, there are two ways of assessing behaviour; locally, or globally. Local behaviour assessment involves examining behaviour at a number of points in state space, and scoring controllers based solely on the behaviour at a number of points. Global behaviour assessment involves running long trials, and scoring controllers based on performance in an extended trial passing through many areas of state space.

Often, as is the case with performing a three point turn, it is beneficial to use a local method of behavioural assessment for the first phase of learning, and a global method for the final phase. It can be seen when learning to perform a three point turn that using a local method will only get so far, and a more global method of assessment needs to be used to "fine tune" the controllers. Using such a method from the start, however, will not work. To use a parallel from human learning, there are many stages in learning to be undertaken before a violinist can perform Brahms' concerto, or even attempt to perform it.

With more complex systems, similar reasoning applies. The role of controller designer becomes more one of tutor, as the right method of assessment has to be chosen. Using today's single stage learning algorithms this problem does not arise, but it is likely that it will be a focus for research in the future.

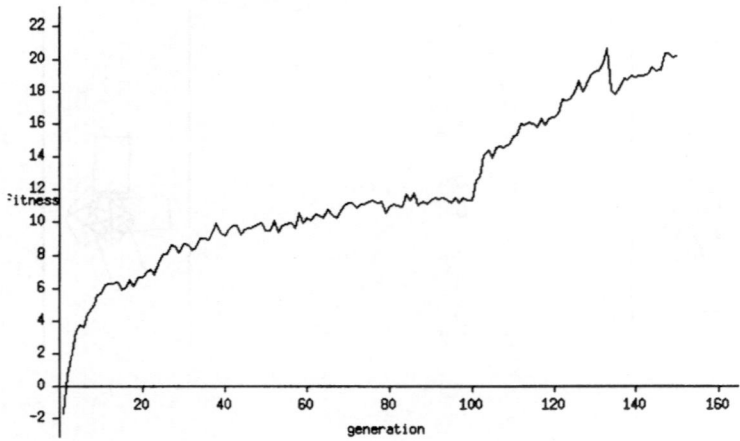

Fig. 12. Convergence rate for two stage learning

6 Conclusion

This paper has described a novel coding scheme for describing fuzzy rule bases for use in genetic algorithms, although the coding scheme could easily be extended to describe a great variety of systems. We have also described a method of minimising epistasis, and thus maximising the efficiency of the coding using a very simple method which utilises the power of the GA to improve its own performance.

The coding scheme leads to enhanced convergence when applied to simple control problems and compared to other similar schemes, although if a rule structure is assumed in advance the performance of the controllers produced can be beaten by conventional schemes. Such a structure, however, could not realistically be determined for more complex problems. We have examined the application of our algorithm to more demanding tasks, and shown how it will produce AGV controllers capable of simple tracking tasks. For more advanced behaviours, the simple approach described is insufficient, and we have described a method of avoiding the pitfalls associated with increasing complexity, and give a successful example of its implementation.

6.1 Further work

Although the coding scheme and algorithm have produced some interesting results, the problems solved have all been rather simple. We are at present working on applying the algorithm to more complicated behaviours, such as parallel parking in a space of varying size and position, without the use of external markers to indicate position.

The possibility of automatic generation of intermediate level behaviours is being considered, as although for tasks such as the multi point turn, behavioural decomposition is easy, for something more involved decomposition may not be so easy. The power of the GA is in producing emergent effects from a number of simple behaviours, and it is often very difficult even to predict these, far less design them, so it seems likely that an automatic process for generating and testing intermediate levels of behaviour will be necessary if the algorithm is to be applied further, particularly if it is to be used for designing dynamic systems that have internal states and consequently have a greatly enhanced level of behavioural complexity.

References

1. Altenberg, L.; *The Evolution of Evolvability* Ch. 3 in Ed. Kinnear, K. E.; *Advances in Genetic Programming*; MIT Press, Cambridge, MA, 1994.
2. Angeline, P. J.; *Genetic Programming and Emergent Intelligence* Ch. 4 in Ed. Kinnear, K. E.; *Advances in Genetic Programming*; MIT Press, Cambridge, MA, 1994.
3. Beasley, D., Bull, D. R. and Martin, R. R.; *An Overview of Genetic Algorithms, Part 1, Fundamentals*; University Computing, Vol. 15, No. 2, 1993.
4. Brooks, R.; *A Robust Layered Control System for a Mobile Robot*; IEEE Trans. Robotics and Automation, Vol. 2, No. 1, Mar. 1986.
5. Cooper, M. G. and Vidal, J. J.; *Genetic Design of Fuzzy Controllers*; Proc. 2nd Int. Conf. on Fuzzy Theory and Technology, Durham, NC, 1993.
6. Davidor, Y.; *Genetic Algorithms and Robotics*; World Scientific, Singapore, 1991.
7. Goldberg, D. E.; *Genetic Algorithms in Search, Optimisation and Machine Learning*; Addison-Wesley, 1989.
8. Hoffmann, F. and Pfister, G.; *Automatic Design of Hierarchical Fuzzy Controllers Using Genetic Algorithms*; Proc. 2nd European Congress on Intelligent Techniques and Soft Computing (EUFIT '94), Aachen, Germany, 1994.
9. Holland, J. H.; *Adaptation in Natural and Artificial Systems (2nd ed.)*; MIT Press, Cambridge, MA, 1992.
10. Karr, C. L.; *Design of a Cart-Pole Balancing Fuzzy Logic Controller using a Genetic Algorithm*; SPIE Conf. on Applications of Artificial Intelligence, Bellingham, WA, 1991.
11. Kosko, B.; *Neural Networks and Fuzzy Systems*; Prentice Hall, Englewood Cliffs, NJ, 1992.
12. Leitch, D. and Probert, P.; *Context Dependent Coding in Genetic Algorithms for the Design of Fuzzy Systems*; Proc. IEEE/Nagoya University WWW on Fuzzy Logic and Neural Nets / Genetic Algorithms, Nagoya, Japan, 1994.
13. Levenick, J. R.; *Inserting Introns Improves Genetic Algorithm Success Rate: Taking a Cue from Biology*; Proc. 4th Int. Conf. on Genetic Algorithms, 1991.
14. Nomura, H., Hayashi, I. and Wakami, N.; *A Self-Tuning Method of Fuzzy Reasoning by Genetic Algorithm*; Proc. Int. Fuzzy Systems and Intelligent Control Conf., Louisville, KY, 1992.
15. Pedrycz, W.; *Fuzzy Sets and Systems*; Research Studies Press, 1989.

16. Takagi, H. and Lee, M.; *Neural Networks and Genetic Algorithm Approaches to Auto-Design of Fuzzy Systems*; Proc. 8th Austrian Artificial Intelligence Conference, FLAI '93, Springer-Verlag, Berlin, 1993.
17. Varšek, A., Urbančič, T. and Filipič, B.; *Genetic Algorithms in Controller Design and Tuning*; IEEE Trans. on Systems, Man and Cybernetics, Vol. 23, No. 5, 1993.
18. Wang, L. X. and Mendel, J. M.; *Generating Fuzzy Rules by Learning from Examples*; IEEE Trans. Systems, Man and Cybernetics, Vol. 22, No. 6, 1992.
19. Zadeh, L.; *Fuzzy Sets*; J Information and Control, Vol. 8, pp. 338 - 353, 1965.

A New Approach to Genetic Based Machine Learning and an Efficient Finding of Fuzzy Rules
— Proposal of Nagoya Approach —

T. Furuhashi, Y. Miyata, K. Nakaoka, Y. Uchikawa

Dept. of Information Electronics, Nagoya University
Furo-cho, Chikusa-ku, Nagoya 464-01, Japan
Tel.+81-52-789-2792
Fax.+81-52-789-3166
E-mail furuhashi@nuec.nagoya-u.ac.jp

Abstracts: This paper presents a new approach to genetic-based machine learning (GBML). The new approach utilizes mechanisms of genetic recombination in bacterial genetics, and the authors have called the new approach "Nagoya approach". The Nagoya approach is efficient in improving local portions of chromosomes. An obstacle avoidance problem for a mobile robot is simulated using the Nagoya approach, and complex fuzzy rules are found.

1. Introduction

Fuzzy inference described in linguistic IF-THEN rules has been widely used for its high performance in human-computer interactions. The difficulties associated with fuzzy inference involve the acquisition of fuzzy rules and the tuning of membership functions. To address these problems, the genetic algorithm (GA)[1], one of the basic models of evolution and an effective tool for constructing evolvable/adaptive complex systems, has been applied[2-7]. C. L. Karr[2, 3] has proposed the application of the GA to the design of fuzzy logic controllers, and his work was a pioneering effort in the application of the GA to fuzzy controls. M. Valenzuela-Rendon[4] has proposed a fuzzy classifier system (FCS) by introducing fuzzy logic into the classifier system[8] and applying the FCS to approximate a nonlinear function. T. Furuhashi et al.[5-7] have studied the application of the FCS to knowledge finding for fuzzy controls. M. A. Lee and H. Takagi[9] have devised another interesting approach to the fusion of fuzzy logic and the GA. They have presented a method to control the parameters of the GA, i.e. mutation rate, crossover rate, etc., by fuzzy logic. In this paper, the authors describe an efficient method for finding fuzzy rules using a genetic-based machine learning (GBML) technique.

There are two distinct approaches to the GBML, called the Michigan approach[8] and the Pitt approach[10]. The Michigan approach uses a single set of production rules or classifiers. Each rule in the population represents a single rule, and the entire population is used to represent the complete rule base. The production system, utilizing this set of rules, senses the environment and outputs actions. Each individual rule has a strength which indicates the utility of the rule in achieving the goal of the system. The apportionment of credit system obtains payoffs from the task environment for the actions and shifts the strength of the rules in proportion to the amount of payoffs as well as to the contribution of the rules to the goal. Genetic operators are applied to the rules on the basis of strength.

An individual in the Pitt approach comprises a set of rules. There are sets of rules in the population; each set of rules is used in the production system, and the payoffs from the environment are directly given to the set of rules. The fitness value of each individual set of rules is assigned in proportion to the amount of payoffs. The Pitt approach does not need the apportionment of credit system.

Through some experiments using the Pitt approach, it has been found that the improvement of the individual rules of a chromosome is hard to achieve. This is because the payoff from the environment combines all the information: the task level performance, the number of rules that fire, the degrees of contribution of rules, etc. into a scalar value. J. J. Grefenstette has proposed a multilevel credit assignment[11] in the context of the Pitt approach to assign credit at the level of individual rules using the bucket brigade algorithm.

This paper presents a new approach to the GBML which is simple and very efficient in improving individual production rules within the total set of rules. We introduce mechanisms of genetic recombination in bacterial genetics[12] into the GBML. The authors call the new approach "Nagoya approach". In this study, the new approach is applied to find fuzzy rules for an obstacle avoidance problem involving a mobile robot. The simulation result shows that the new GBML method is very efficient in finding control rules with limited information from the payoffs provided by the task environment.

2. Pitt Approach

Figure 1 shows an example of flow of the Pitt approach applied to knowledge finding for obstacle avoidance. The initial population is randomly generated. There are n_{chr} chromosomes, each of which consists of n_r rules. Each chromosome is used to steer the robot to avoid a moving obstacle. The payoffs are given to each chromosome based on its performance. All the chromosomes are tested one by one and their fitness values are determined in proportion to their own payoffs. The stopping condition is, for example, the achievement of all the tasks by at least one chromosome. If the stopping condition is not met, the rule generation mechanism is applied to the population. The genetic operators such as selection, reproduction, crossover and mutation are applied to the chromosomes to generate new rule sets. Through the iteration of the trials of obstacle avoidance by the chromosomes and the generation of new chromosomes, the tasks will finally be achieved.

It has been found through some experiments that it is difficult to focus improvement on the individual rules in a chromosome by the Pitt approach, because the payoffs from the environment to a chromosome are scalar values assigned to its performance.

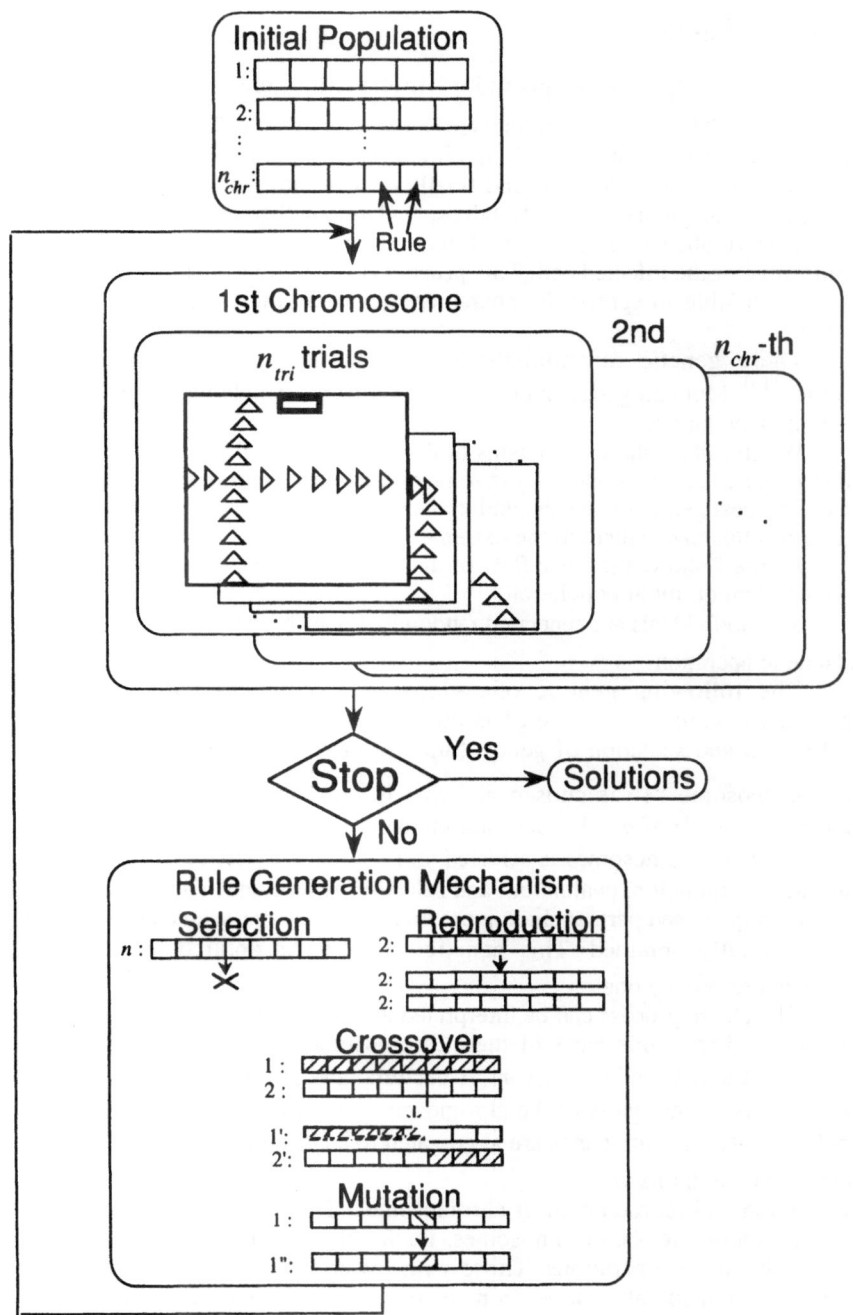

Fig.1 Flow of Pitt approach

3. Nagoya Approach

Bacterial genetics provides interesting mechanisms for genetic recombination[12]. Bacteria can transfer DNA to recipient cells through mating. Male cells transfer strands of genes into female cells. Then the female cells acquire characteristics of the male cells, and finally change into male cells. By these means, the characteristics of one bacteria can be spread among the entire bacteria population.

Bacteriophages carry a copy of the host gene across and incorporate it into the chromosome of the infected cell. This process is called transduction. By transduction, it is also possible to spread the characteristics of a single bacterium among other bacteria.

These genetic recombinations have led to a mechanism of microbial evolution[13]. Mutated genes can be transferred from a single bacterium to others and effect rapid evolution.

We introduce the mechanisms of the above bacterial genetics into the GBML. Multiple bacteria are reproduced and some genes of each chromosome are mutated and tested. The best genes are chosen and transferred to other bacteria. The new approach streamlines this mechanism in the extreme.

Figure 2 shows the basic flow of the Nagoya approach.

(a) Generation of initial population

n_{chr} individuals are generated randomly. Each chromosome is evaluated.

(b) Genetic operation

The following genetic operations are applied to individuals and new populations of chromosomes are generated:

(i) Mutation and selection of genes: Suppose there are n_p parts in a chromosome.

One chromosome "1" is chosen and this is reproduced to m clones. Part i (i is randomly decided) of m - 1 clones are mutated. Each chromosome is evaluated. The elite among m chromosomes is selected and the rest are deleted. The above process-reproduction, mutation, evaluation, and selection-is repeated. The mutation is applied to a randomly chosen part, excluding previously chosen parts, and a new chromosome "1' " is finally obtained. This genetic operation is applied to all the n_{chr} chromosomes one by one.

The above process can be interpreted as follows: A bacterium is reproduced to m clones and the same parts of their chromosomes are mutated. The elite part is selected and transferred to other m - 1 bacteria, and inferior genes are replaced with elite one. Then other parts of the chromosomes are mutated, evaluated, selected and transferred. All the elite genes are aggregated to a chromosome of one bacteria. n_{chr} bacteria evolve in this way.

(ii)Selection and reproduction of chromosomes: The selection and the reproduction steps are applied to the chromosomes. Each chromosome has its fitness evaluated through the above operations. Those with lower fitness values are deleted. Some chromosomes randomly chosen from the remaining chromosomes are reproduced. In Figure 2, chromosome k' is deleted, since its fitness value is the least. The chromosomes i', j' are selected and reproduced.

177

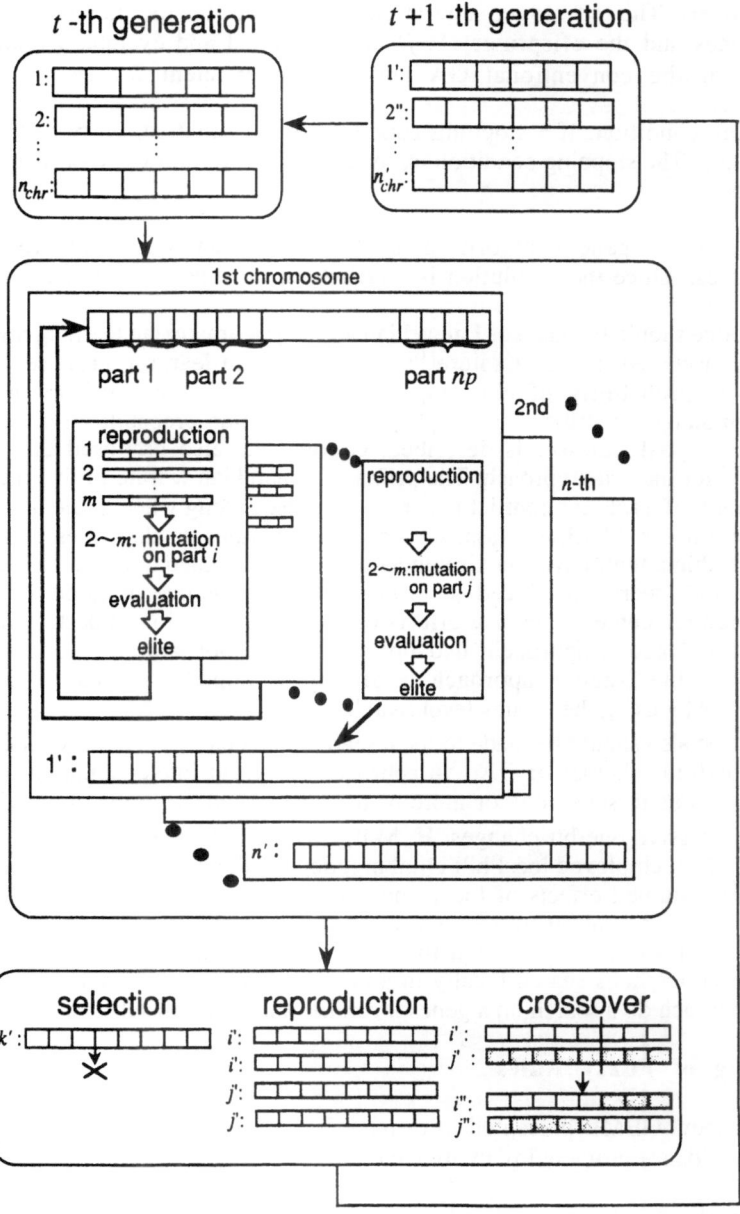

Fig.2 Flow of Nagoya approach

(iii)Crossover: The crossover operation is applied to the newly generated chromosomes and the offsprings i'', j'' are generated and evaluated. This is an operation of the conventional GA and is also efficient for improving the chromosomes.

(iv)Stopping condition: If a stopping condition is not satisfied, go back to (ii). If satisfied, stop. The stopping condition is, for example, a certain value or a number of generations.

The above genetic algorithm is efficient in the local improvement of chromosomes, since the evolution is carried out on the level of chromosomal genes.

Notice should be made of Lamackian learning. Lamarckian learning algorithm have been discussed in the literature[14-16]. Lamarckian learning proposed in [14] changes the probability of choosing crossover points according to the error distribution along the string. Also in [15], the Lamarckian probability for mutations based on sub-goal fitnesses is described. Grefenstette also proposed Lamarckian learning[16] for the Pitt Approach. The primary Lamarckian feature of this method is that a strength of a rule for conflict resolution is passed along when a rule is inherited to an offspring. In addition, crossover first clusters rules so that rules that fire in sequence within a high-payoff environment tend to be assigned to the same offspring. These methods were introduced to promote the inheritance of good strings by diminishing the counterproductive effects of crossover/mutation in destroying good schema. The Nagoya approach also promotes the inheritance of good schema. Furthermore, the Nagoya approach is aiming at improving local portions of chromosomes by using the "genes-level search."

Notice also should be made to local search methods[17-20]. D. E. Goldberg [17] described hybrid schemes of GA. An approach to hybrid implementation for local optimization was to select one or more of the best strings from the population and perform successive one-bit changes. H. Mühlenbein[18] proposed a parallel genetic algorithm which employed local hill climbing, done with a simple 2-opt exchange. P. Jog, et al.[19] studied effects of local improvement operators specially devised for solving the traveling salesman problem. J. A. Miller, et al.[20] also discussed several local improvement operators which flip binary code in each locus on/off for local search. These methods search locally in a neighborhood, i.e. bit by bit sweep. The Nagoya approach do a search on a genes level, i.e. mutations on substrings.

4. Finding of Fuzzy Rules

The new GBML presented in chapter 3 is applied to find fuzzy rules for a collision avoidance problem involving a mobile robot.

4.1 Coding

The paper uses a method of coding of knowledge into a gene as shown in Fig. 3. A kind of knowledge concerned with the input x_1 is expressed by

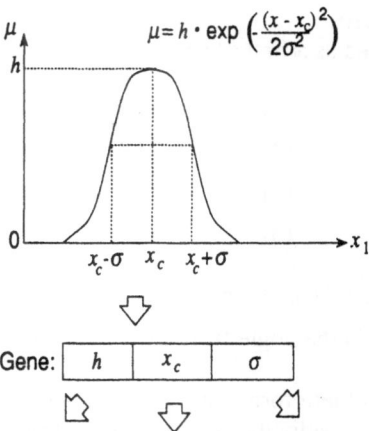

Gene:

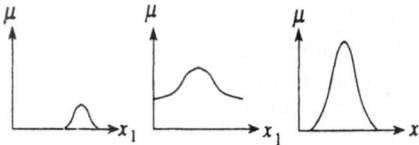

Various knowledge can be expressed.

Fig.3 Example of Coding of knowledge into a gene

$$\mu = h \bullet \exp\left(-\frac{(x_1 - x_c)^2}{2\sigma^2}\right) \qquad (1)$$

where x_c is the central position, σ is the width (standard deviation) and h is the height of this function. Using these three parameters as shown in the figure, various types of knowledge can be expressed. The three parameters are encoded into the gene. This knowledge representation has a clear correspondence with a membership function employed in fuzzy logic. The knowledge defined by this function can be labeled and corresponds to a symbolic expression. Even if the function does not resemble the shape of a conventional membership function, the knowledge can still be handled and some inference can be carried out.

Figure 4 shows an example of if ~ then ~ rules encoded into a chromosome. The figure shows the case where there are two inputs x_1, x_2 and one output y and n_r rules in a part. Each rule has the parameters of eq. (1), i.e. the heights h_1, h_2, the central positions x_{c1}, x_{c2}, and the widths σ_1, σ_2 for the inputs x_1, x_2, respectively. For the output y, the central position x_y and the width σ_y are encoded. For clarity and simplicity, the height h_y is fixed at unity. The parameter $t \in [0, \ 1]$ defines the truth value in the antecedent in between the product-sum-center of gravity and the sum-sum-center of gravity.

4.2 Simulation conditions
A robot is approximated as a first order system given by

$$T \, d\omega / dt + \omega = u \qquad (2)$$

$$\theta = \int \omega \, dt + \theta_0 \qquad (3)$$

$$dx / dt = |V| \sin \theta \qquad (4)$$

$$dy / dt = |V| \cos \theta \qquad (5)$$

where T is the time constant, $|V|$ is the speed of the robot, u is the steering angle, ω is the angular velocity, θ is the angle from the north, and x, y are the coordinates of the robot.

Figure 5 shows the problem formulation. The robot starts from a line opposite to the goal. The initial angular velocity of the robot is 0 rad/s. It is a success when the robot reaches the goal, and a failure when the robot collides with a moving obstacle or goes off the screen. The faster the robot reaches the goal, the better the evaluation. The nearer the failed point, also the better the evaluation.

Figure 6 shows the denotations of relative distances and velocities between the robot and the moving obstacle as well as the goal. The speeds of the robot and the moving obstacle are $|V|$, $|V_0|$, respectively. The speed of robot will be changed by

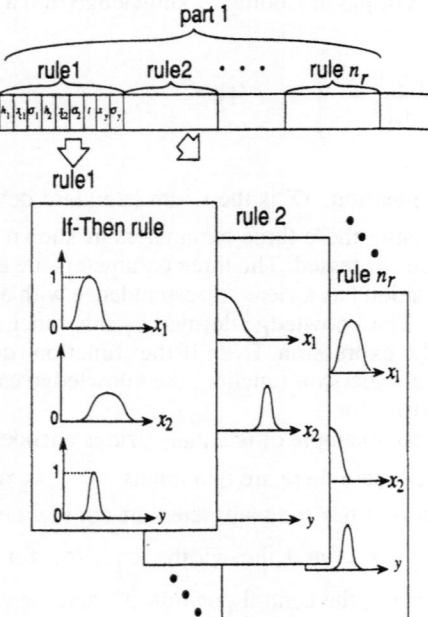

Fig.4 Example of if ~ then ~rules encoded into a chromosome

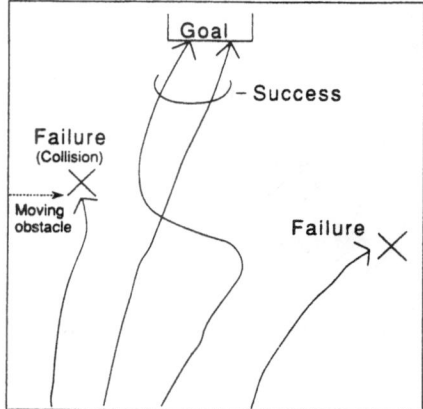

Fig.5 Problem formulation

$$|V| = |V_s| + \Delta V \qquad (6)$$

where $|V_s|$ is a constant value and ΔV is the manipulated variable determined by the control rules. The speed of moving obstacle $|V_0|$ is set to be constant. The distance between the robot and the obstacle is denoted by D. The angle between the direction of the robot and the direction of the obstacle viewed from the robot is denoted by δ. In the same way, the angle between the direction of the robot and the goal is ψ. V_R is the relative velocity between the robot and the obstacle. The angle between the relative velocity V_R and the direction of the robot from the moving obstacle is ϕ.

When ϕ is nearly zero, it means that the moving obstacle comes toward the robot. Each angle is set to be counterclockwise positive viewed from the reference line in Fig. 6.

The control rules used in this paper are expressed in Figure 7. The number of parts in this chromosome is tentatively set at three. Each part has no predetermined particular task in this simulation. Each part has five rules. The inputs of each rule are D, ϕ, δ, ψ, and the angular velocity of the robot ω. The output of the rule is the steering angle u and the change of speed ΔV. The truth value in the antecedent of the rule is determined by t_{ij} ($i = 1, \bullet\bullet\bullet , 5, j = 1,2,3$). If $t_{ij}= 0$, then the product of grades in the antecedent is used. If $t_{ij}= 1$, then the sum is the truth value. The truth value is in between [0, 1] in proportion to t_{ij}. Every parameter is randomly generated. The control rules are evolved by the new GA through the interaction with the environment. The new GA used here has to determine $20 \times 5 \times 3 = 300$ parameters.

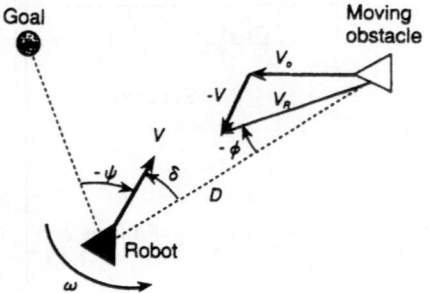

Fig.6   Denotations of parameters

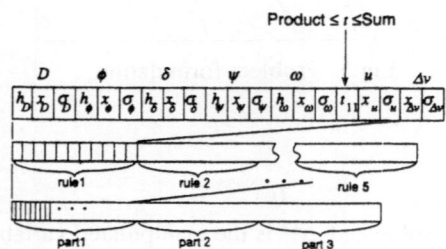

Fig.7   Coding of rules into a chromosome

4.3 Application of new approach

The new GBML is applied in the way shown in Fig. 8. In this simulation, there are ten chromosomes in a population. Each chromosome is reproduced to (n_p=) four. Each rule in a randomly chosen part of the newly generated chromosomes is mutated. One of the 20 parameters of each rule is selected and changed using uniform random numbers. Then each chromosome controls the robot using all the rules in the chromosome simultaneously and is evaluated under five different simulation conditions shown in Fig.9. The robot starts from the central point on the bottom line. The moving obstacle comes from the five different points indicated in the figure. The definitions of payoffs are given by the equations below. In case of success, the gene receives the following payoff:

$$1 + \frac{100}{\text{No. of steps}}. \qquad (7)$$

In case of failure, the gene receives the payoff given by

$$\frac{10}{\text{Distance to the goal}}. \qquad (8)$$

It takes at least 200 steps for the robot to reach the goal from the starting point. There are 640×750 pixels in the working area on the screen. The unit of the distance is this pixel on the screen. The elite gene with the highest payoff among the four genes is selected and transferred to other three chromosomes to replace the inferior genes. Then another part of each chromosome except for the original one is mutated and tested. Through this mutation and transfer operation, a new chromosome 1' is obtained. This genetic operation is applied to the remaining nine chromosomes one by one.

The selection and reproduction operations are applied to the ten newly generated chromosomes. Two of the chromosomes with the least fitness values are deleted. Two out of the remaining eight chromosomes are reproduced and the new chromosomes are changed by the one point crossover operation.

4.4 Results

Figure 10 shows changes of fitness values with the Nagoya approach. The horizontal axis is the generation. The fitness value of the elite chromosome and the average

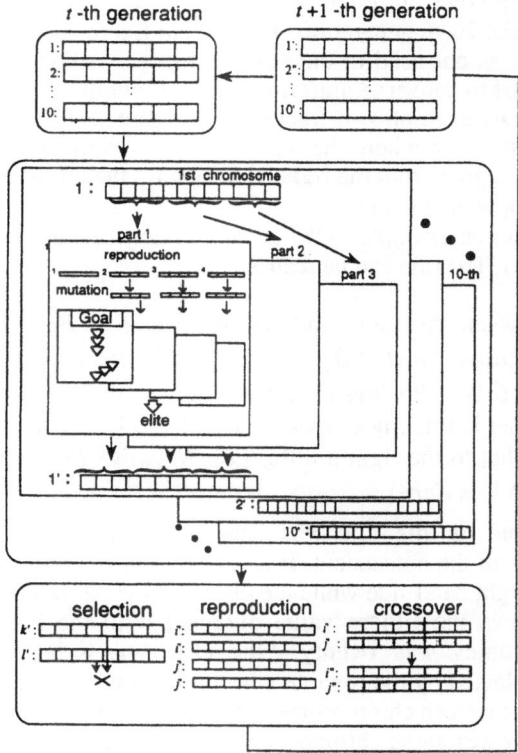

Fig.8 Application of New GBML

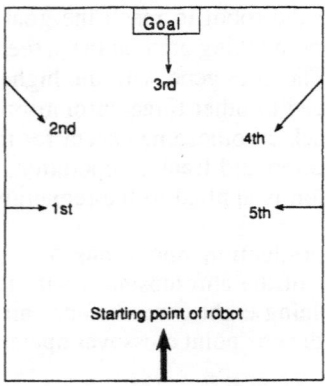

Fig.9   Simulation conditions

fitness value of the ten chromosomes are shown in the figure. The top figure shows a case where the new GBML is applied. The elite chromosome obtained by the new approach succeeded in controlling the robot to reach the goal under all of the five conditions at around the 20-th generation. The bottom figure shows a case where the crossover operation was not used in the new approach. Without the crossover, the chromosomes were fast to converge uniformly to local minima.

Figure 11 shows examples of tracks of the robot and the moving obstacle. Figure 11 (a) shows the case where the obstacle comes from the fore and (b) is the case where the obstacle goes from the right hand side to the left in front of the robot. The tracks of the robot were smooth.

Figure 12 shows an example of the acquired fuzzy rule used in the early stage of control in Fig. 11(a). The rule can be read as:

If the distance between the robot and the obstacle D is Big, the danger of collision is $Negative$ $Small$ (ϕ is $Negative$ Big), the obstacle is in the right hand side (δ is $Positive$ Big), the goal is on the left hand side (ψ is $Negative$), the robot is rotating clockwise a little (ω is $Negative$ $small$),
Then steer the robot to the right a little (u is $Negative$ $Small$) and the robot is not braked (ΔV is $Zero$).

The truth value of this rule was nearly the sum of the grades of the membership functions in the antecedent. It is known that the rule worked simply to steer the robot to the right hand side while the obstacle was far away.

Figure 13 shows the fitness value obtained with the Pitt approach. The number of chromosomes was twenty. One elite was selected first. Fifteen chromosomes were selected by the roulette wheel selection. Four chromosomes were selected from the above sixteen chromosomes and reproduced. The one-point crossover is applied to the newly generated chromosomes. The mutation was applied to all the chromosomes except for the elite one at a rate of 10%. It was difficult to find out a set of rules which succeeded in all the five conditions with the conventional approach even after the 275-th generation. A comparison of the performances of the proposed

185

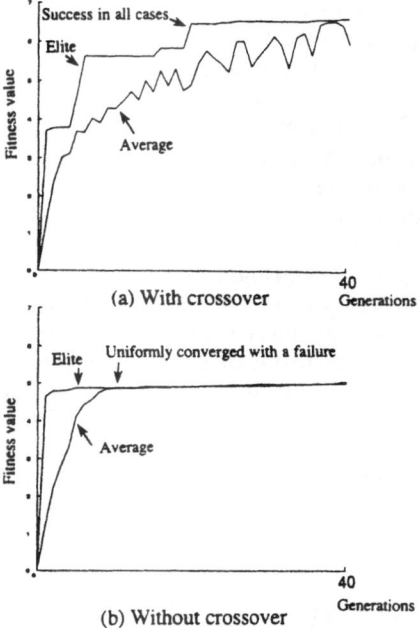

(a) With crossover

(b) Without crossover

Fig.10   Fitness values with Nagoya approach

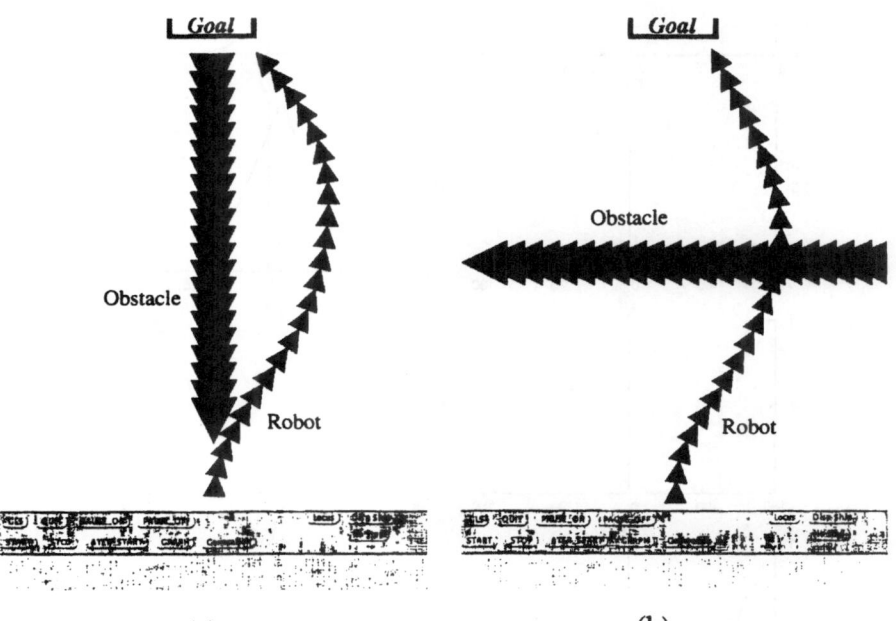

(a)   (b)

Fig.11   Tracks of robot with new approach

approach and the conventional approach can be made on the basis of evaluation times. A chromosome is evaluated under the five different conditions. This evaluation is done less times in a generation by the Pitt approach. The total evaluation times of the Pitt approach in the 275 generations were twice as much as those of the Nagoya approach in the 20 generations. The CPU time of the Pitt approach at the 275-th generation was also two times larger than that of the proposed method at the 20-th generation. It was difficult to obtain a better result with the Pitt approach with different mutation rates/different numbers of populations/etc. Nagoya approach constantly yielded good results with wide ranges of crossover rate/mutation rate/number of clones.

Figure 14 shows the tracks of the robot at 275-th generation obtained by the Pitt approach. Figure 14 (b) shows clearly that the local improvement of the chromosome is barely achieved by the conventional approach.

Figure 15 shows another result of simulations. The simulation in Fig.10(a) (Nagoya approach) was repeated 10 times with different initial chromosomes. The averaged fitness values of elite and average are shown. The simulation in Fig. 13 (Pitt approach) was also repeated 10 times. The number of populations was increased to 100 for this simulation. The averaged fitness values of this simulation is also shown in the figure. The new approach showed a higher performance in achieving the goal.

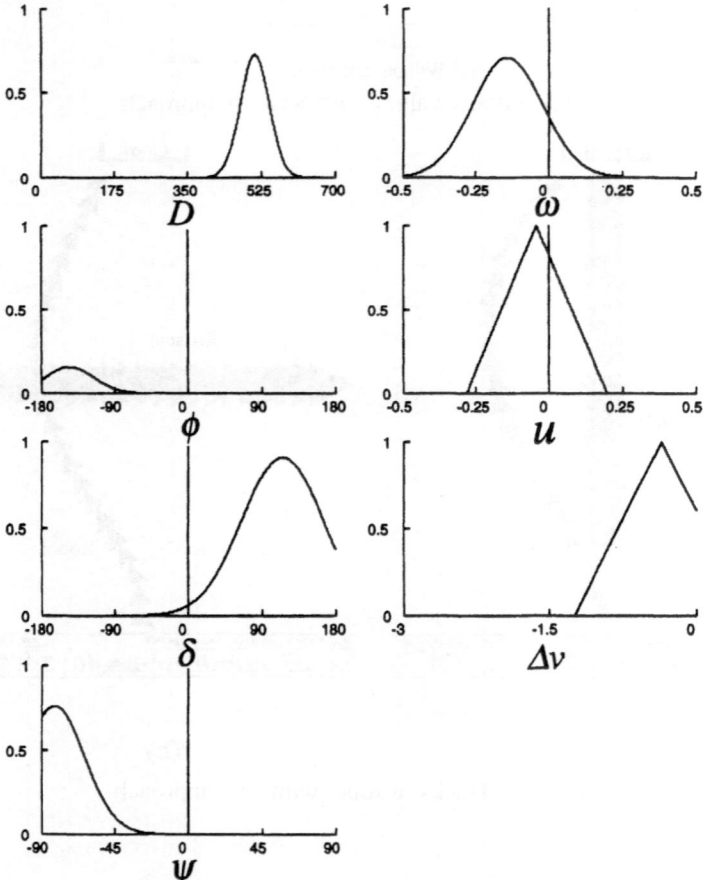

Fig.12 Example of acquired fuzzy rule

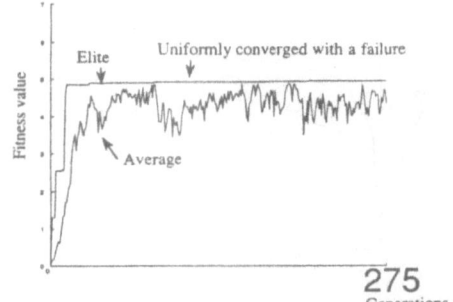

Fig.13   Fitness value with Pitt approach

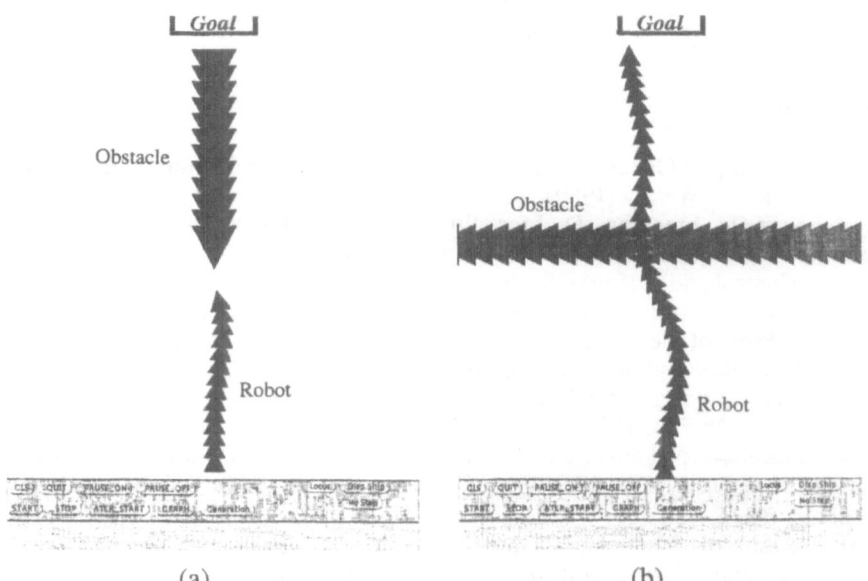

(a)   (b)

Fig.14   Tracks of robot with Pitt approach

5. Conclusions

This paper presented a new GBML called Nagoya approach. The new approach utilizes the mechanisms of genetic recombination in bacterial genetics. The new algorithm is efficient particularly in the local improvement of chromosomes and is expected to be effective for the construction of evolvable/adaptive complex systems. It is expected that the Nagoya approach can be applied to various types of machine learning problems, and its high performance in improving local sets of rules works well to solve complex tasks.

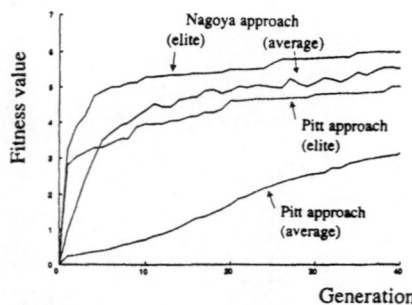

Fig.15 Fitness values with Nagoya/Pitt approach
(Averaged values of 10 trials)

6. References

[1] D.E.Goldberg, "Genetic Algorithm in Search, Optimization and Machine Learning", Addison Wesley (1989)
[2]C. L. Karr, L. Freeman, D. Meredith, "Improved Fuzzy Process Control of Spacecraft Autonomous Rendezvous Using a Genetic Algorithm", SPIE Conf. on Intelligent Control and Adaptive Systems, pp.274-283, 1989
[3] C. L. Karr, "Design of an Adaptive Fuzzy Logic Controller Using a Genetic Algorithm", Proc. of the 4th Int'l Conf on Genetic Algorithms, pp.450-457, 1991
[4] M. Valenzuela-Rendon, "The Fuzzy Classifier System: A Classifier System for Continuously Varying Variables", Proc. of the 4th Int'l Conf on Genetic Algorithms, pp.346-353, 1991
[5]T. Furuhashi, K. Nakaoka, K. Morikawa, Y. Uchikawa, "Controlling Excessive Fuzziness in a Fuzzy Classifier System", Proc. of the 5th Int'l Conf on Genetic Algorithms, p.635, 1993
[6]T. Furuhashi, K. Nakaoka, K. Morikawa, Y. Uchikawa, "An Acquisition of Control Knowledge Using Multiple Fuzzy Classifier Systems", Journal of Japan Society for Fuzzy Theory and Systems, Vol.6, No.3, pp.603-609, 1994
[7] K. Nakaoka, T. Furuhashi, Y. Uchikawa, "A Study on Apportionment of Credits of Fuzzy Classifier Systems for Knowledge Acquisition of Large Scale Systems", Proc. of the 3rd Int'l Conf. on Fuzzy Systems, pp.1797-1800, 1994
[8] J. H. Holland, J. S. Reitman, "Cognitive Systems Based on Adaptive Algorithms", in *Pattern Directed Inference Systems*, D. A. Waterman, F. Hayes-Roth (Eds.), pp.313-329. Academic Press, New York, 1978
[9] M. A. Lee, H. Takagi, "Dynamic Control of Genetic Algorithms Using Fuzzy Logic Techniques", Proc. of the 5th Int'l Conf on Genetic Algorithms, p.76-83, 1993
[10] S. F. Smith,"A Learning System Based on Genetic Adaptive Algorithms", Ph. D. Thesis, University of Pittsburgh, 1980
[11] J. J. Grefenstette, "Multilevel Credit Assignment in a Genetic Learning System", Proc. of the 2nd Int'l Conf. on Genetic Algorithms, pp.202-207, 1987

[12] R. Schleif, "Genetics and Molecular Biology (2nd Ed.)", The Johns Hopkins Univ. Press, 1993

[13] L. Margulis, D. Sagan, "Microcosmos -- Four Billion Years of Microbial Evolution", Summit Books, 1986

[14] J. D. Schaffer, A. Morishima, "An Adaptive Crossover Distribution Mechanism for Genetic Algorithms", Proc. of the 2nd Int'l Conf. on Genetic Algorithms, pp.36-40, 1987

[15] Y. Davidor, "A Genetic Algorithm Applied to Robot Trajectory Generation", in Handbook of Genetic Algorithm, L.Davis, Ed., Van Nostrand Reinhold, ch.12, 1991

[16] J. J. Grefenstette, "Lamarckian Learning in Multi-agent Environments", Proc. of the 4th Int'l Conf. on Genetic Algorithms, pp.303-310, 1991

[17] D.E.Goldberg, "Genetic Algorithm in Search, Optimization and Machine Learning", pp.202-204 in Chapter 5, Addison Wesley (1989)

[18] H. Mühlenbein, "Parallel Genetic algorithms, Population Genetics and Combinatorial Optimization", Proc. of the 3rd Int'l Conf. on Genetic Algorithms, pp.416-421, 1989

[19] P. Jog, J. Y. Suh, D. V. Gucht, "The Effects of Population Size, Heuristic Crossover and Local Improvement on a Genetic Algorithm for the Traveling Salesman problem", Proc. of the 3rd Int'l Conf. on Genetic Algorithm, pp.110-115, 1989

[20] J. A. Miller, W. D. Potter, R.V. Gandham, and C. N. Lapena, "An Evaluation of Local Improvement Operators for Genetic Algorithms", IEEE Trans. on Systems, Man, and Cybernetics, Vol.23, No.5, 1993

A Neuro-Money Recognition Using Optimized Masks by GA

*Fumiaki Takeda, **Sigeru Omatu

*GLORY LTD. Development Center 3-1 Shimoteno 1-Chome, Himeji, 670, Japan
Tel:+81-792-92-8455 Fax:+81-792-94-9603

**University of Osaka Prefecture, College of Engineering, Department of Computer and
systems sciences, Sakai Gakuen Cho 1-1, Osaka Prefecture 593, Japan
Tel:+81-722-52-1161(2288) Fax:+81-722-59-3340 E-mail omatu@cs.osakafu-u.ac.jp

Abstract: Up to now, much research of the application to neural networks (NN) has been
reported. We have proposed a neuro-pattern recognition for bill money with masks and have
reported its effectiveness for money recognition. Recently, genetic algorithm (GA) is reported
as the effective optimizing method. In this paper, we adopt the GA to mask optimization in
the recognition method. Namely, we regard the position of the masked part in the input
image as a gene. We operate crossover, selection, and mutation to some genes. By
repeating a series of these operations, we can get effective masks for paper currency
recognition. We compare the ability of NN using the optimized masks by the GA with the
one of NN using the random masks determined by random numbers. Then we show that the
GA is effective to optimize masks for the method of neuro-pattern recognition with masks.
Furthermore, we develop high-speed neuro-recognition board to realize the neuro-pattern
recognition for paper currency in the commercial products.

1. Introduction

Up to now, we have proposed money recognition methods by a small size
neural network (NN) to aim at implementation of the NNs for the money recognition
machines [1]-[4]. Especially, we regard the sum of input pixels as a characteristic
value in the proposed money recognition method using a neural network, which is a
called neuro-money recognition method in what follows. This is based on slab
architecture of Widrow's algorithm [5] which is invariant to various fluctuations of
input image. Namely, in our neuro-money recognition method, we have adopted
random masks determined by random numbers in a pre-processor, which means that
two-dimensional input image is measured from various directions [2]-[5]. It enables
us to input the characteristic values to the NN. The sum of non-masked pixels by
random masks is called as slab value. We input not pixel values but slab values to the
NN. We have shown that the recognition of worn out US dollars can be realized by a
small size NN using random masks in the experiment [3][4]. However, in this
method, we must decide a masked area by random numbers. So we may not get
effective masks which reflect the difference of input image to slab values. We need
some optimization for random masks. Under this background, much research on
optimization by the genetic algorithm (GA) has been reported [6].

In this paper, in order to sample the characteristic values of input image effectively, we adopt the GA to the mask optimization in the neuro-money recognition method using random masks. This is a unique technique which applies evolutionary mechanism of a life to the mask optimization. The proposed method on the mask optimization can generate effective masks which satisfy purposive generalization of the NN. We regard a position of a masked part as a gene. We operate "crossover", "selection", and "mutation" for some genes [6]-[8]. By repeating a series of operations, we can get effective masks for money recognition. First, we explain the basic idea of the mask method for the NN [2]-[5]. Second, we discuss the experimental results with random masks using US dollars. Then we show the effectiveness of the optimization method of masks using the GA compared with the random masks in the experiment. Finally, we refer to a high speed neuro-recognition board which will be applied as an input to NNs in commercial products. This neuro-board has a Digital Signal Processor (DSP), which has been developed to commercialize the money-recognition machines by NNs.

2. Basic Idea of Neuro-Money Recognition with Masks

In the present method, image of bill money is assumed to be shown in Fig.1. We can get 31 as the slab value of "$1" in Fig.1(a) and 25 as that of "$5" in Fig.1(b). Here, the slab value has been defined in [2]-[5]. Using them, we can separate "$1" and "$5". However, it may generate the same slab value even when the inputs are different. For example, slab value corresponding to "$10" in Fig.2(a) is 23 but one corresponding to "$20" in Fig.2(b) is also 23. This problem can be solved by adopting a mask which covers some parts of the input in Fig.3(c). We can get 13 as the slab value when the input in Fig.2(a) is covered with the mask in Fig.3(c).

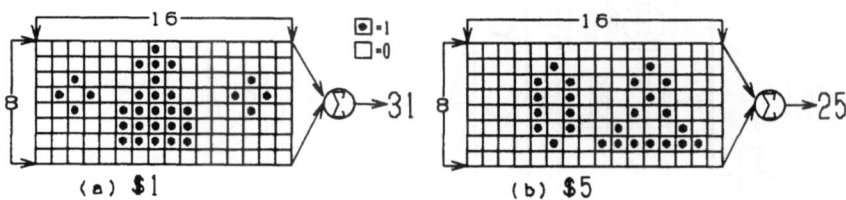

Fig.1. Different input patterns and different slab values.

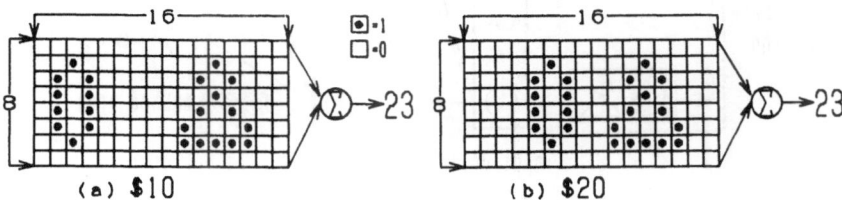

Fig.2. Different input patterns and same slab values

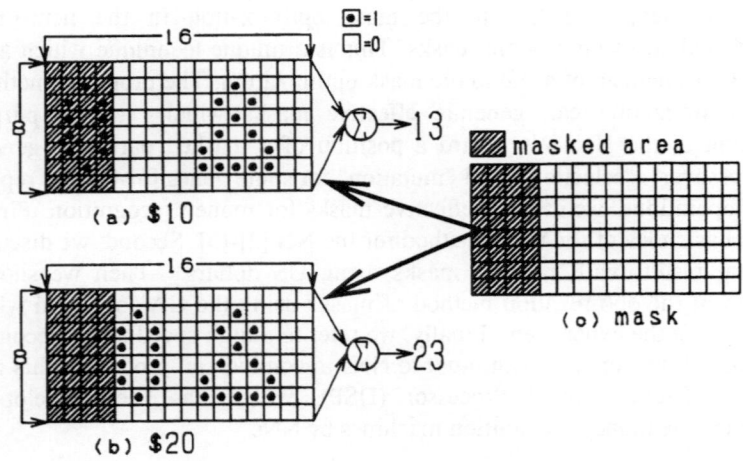

Fig.3. Mask and slab values.

Otherwise, by the same method we can get 23 as the slab value from the input in Fig.3(b). In this way, we can separate "$10" and "$20". Using the masks, we can measure a two-dimensional image from various view-points as if we measure a three-dimensional object. Furthermore, we have shown that learning converges fast since the probability to produce the characteristic slab values becomes high [2][3]. We show the configuration of the proposed NN with masks in Fig.4. Some parts of input are covered with various masks in pre-processing. The sum of pixels which are not masked becomes one slab value which is taken as an input of NN. We select 16 as

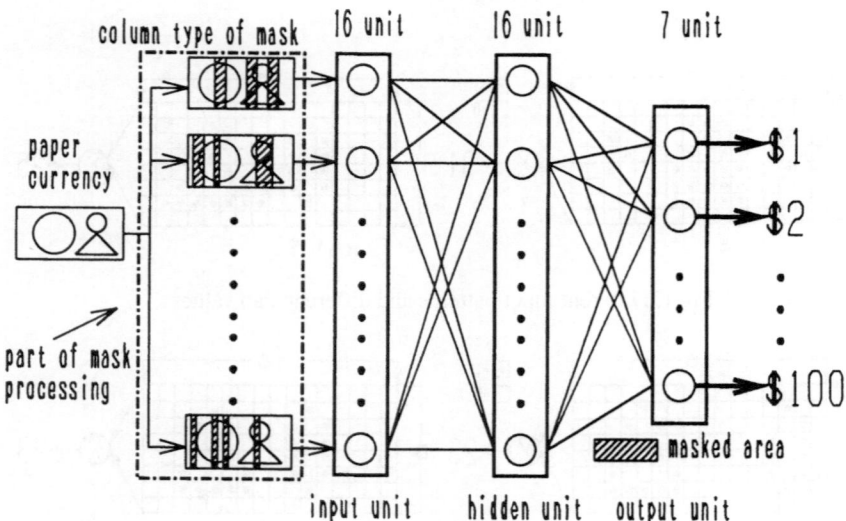

Fig.4. NN construction of the mask method.

the number of masks from the various kinds of simulation [2]-[4]. In the following, we describe 16 masks as a mask set. Both numbers of input units and the hidden ones are 16. The kinds of bill money are US $1, $2, $5, $10, $20, $50, and $100 [3][4]. Thus, the number of output units which corresponds to money kinds becomes 7.

3. Experiment Using Random Masks

Here, we explain the determining method of masks by random numbers. In the following, we describe the mask determined by the random numbers as the random mask. As shown in Fig.5, M is the number of the least masked area (column) and each masked area is ordered. We generate 16 random values among [-1,1] and they are equal to columns number on the input image. We mask the column whose number is equal to ordered number of random values which is minus value. For example, M is 16 and the number of random values is 16 which are generated among [-1,1]. If ordered random value is -0.6 and its ordered number is one, the first column is masked. In this way, we change the initial values which generate random values and repeat their procedure 16 times. Then we determine one kind of masks as shown in Fig.6. We determine 16 non-duplicated masks which are described as mask set. In the following, we generate 30 random mask sets and investigate the generalization of the NN using the unknown data of US dollars which include damaged bill money and fluctuation error by conveying.

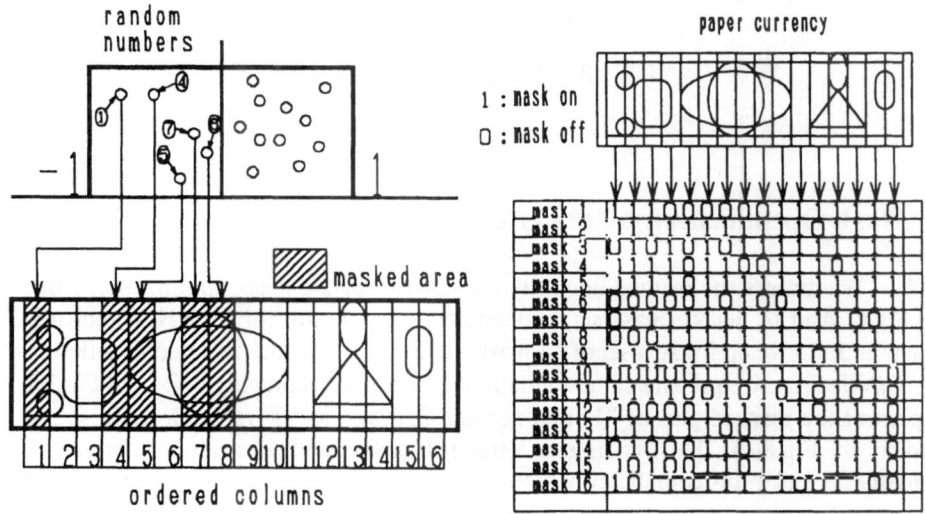

Fig.5. Determination of a random mask set.   Fig.6. A mask set for NN.

In the experiment, we construct the experimental system using the current recognition machine. It can sample image data such as 216×30 pixels which are represented by 1 byte gray level. Bill money is conveyed in the parallel direction of its short part and its conveyed speed is more than 10 pieces per second. Here, one column (the least masked area) denotes a series of 30 pixels in the parallel direction of short part of bill money in this sensor system. We adopt the error back-propagation method with oscillation term for learning [1]-[4]. We use 10 pieces of bill money as learning data. We define one iteration means learning from $1 to $100. We continue learning until iteration number reaches 5,000 times. To evaluate the method, other 30 pieces of bill money for each kind are used.

From the experimental results using 7 kinds of US dollars as shown in Fig.7, generalization of the NN obtained by the 30 random mask sets is from 59% to 99%. In this way, generalization of NN is influenced by mask sets. When we use random search procedure to find the superior mask sets, we must try to investigate at least 30 times like this experiment. This is not effective procedure.

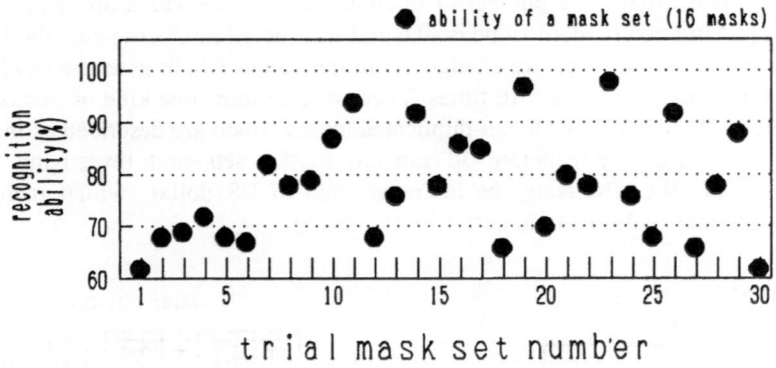

Fig.7. Ability of random mask sets.

4. Mask Optimization by the GA

We discuss the method adopting the GA [6] for the mask optimization based on Fig.8. And 16 masks (mask set) generate one input vector for the NN as shown in Chapter 3. So we optimize a set of 16 masks. The GA procedure is stated as follows:
CODING: First, we prepare some random mask sets (for example, A, B, and C). We represent the masked part as "1" and non-masked part as "0" for each mask set, as shown in Fig.8(a). This coding is obtained easily and satisfies completeness, soundness, and non-redundancy which are proposed as an evaluation standard of coding [9].
CROSSOVER: We crossover their parts (for example, 1/2) of genes in these two parental mask sets as shown in Fig.8(b). While we sample the parental mask sets according to their ability. Namely, we consider the ability of the GA and show that the mask set which has the higher evaluation is sampled easily [9]. Still more, in order to emphasis the effectiveness of this sample method, we adopt the scaling

(New evaluation = evaluation x evaluation). The crossover satisfies character preservingness which is proposed as an evaluation standard of crossover [9]. However, we decide the candidates for the crossover in the 16 masks (mask set) using the random numbers, every each time. In the concrete, we play dice which has the 16 faces and decides the candidate. If the face which is the same as the last one appears, we continue to play the dice till the face which is different from the last one appears.

MUTATION: Furthermore, as shown in Fig.8(c), to provide the variety to crossovered mask set, mutation which reverses the some bits of the genes in the mask is randomly operated during the determination of the new mask. However, we decide the mutation chance and the candidate for the mutation in the 16 masks and the reversed bits using the random numbers, every each time. Then we can obtain inputs of the NN using the mask set.

Learning is executed by using the above inputs. After learning, we investigate the generalization of each NN with mask set using unknown data.

SELECTION: If we select only the descendant mask sets which satisfy target ability, there is some risk such that descendant mask sets will disappear in a few generation. So we must maintain the number of descendant mask sets which are sampled for crossover in the next generation. As shown in Fig.8(d), we replace the parental mask set by the crossovered one when the ability of the crossovered mask set is better than that of the parental one. So the number of descendant mask sets is maintained.

By repeating a series of these operations, we can get the superior mask sets. These mask sets enable us to shorten the learning time and to improve the generalization of the NN. We make an experiment of the mask optimization by the GA. The experimental condition is the same as Chapter 3. The number of mask set is 10. We continue the GA operations till the target mask set whose ability is more than 95% is obtained. Still more, we make the GA experiment three times by changing initial mask sets. But for the NN learning, the initial values of neuron's weights are the same for the each experiment in order to fix the characteristic of the NN even if the mask set are changed. Figure 9 shows the transition of the NN's generalization with mask set (ability of the mask set) by the GA operations.

Figure9 (a) shows the result such that the target mask set is determined in the fifth generation. While Fig.9 (b) shows the result such that the target mask set is not obtained till final generation. However, in this case, we can confirm that it is determined that the mask set whose ability is higher than that of the initial one. Still more, Fig.9 (c) shows the result such that there is the initial mask set which has the target ability and the result such as the transition of other mask sets by continuing the GA operations. In the three experiments, the ability of the sixth mask set as shown in Fig.9 (c) is improved extremely and its improvement is from 64.3% to 87.2%. Furthermore, we can confirm that the average ability of every 10 mask sets is increased gradually for each experiment. From these experiments, we can get some superior mask sets in a short period using the proposed GA operations as compared with the random search procedure. Because the possibility such that the optimized mask set by the GA covers the area which does not have a picture like a watermark is decreased. Furthermore, when we suppose the method such as investigation for the

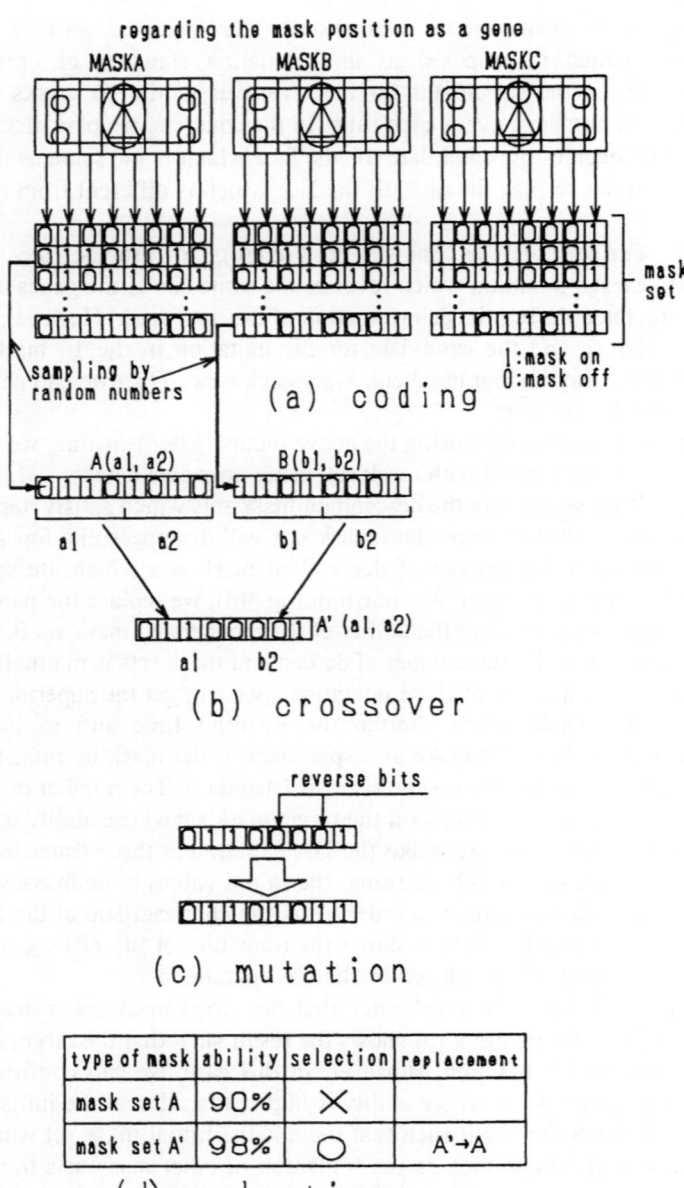

Fig.8. Basic operation of the GA.

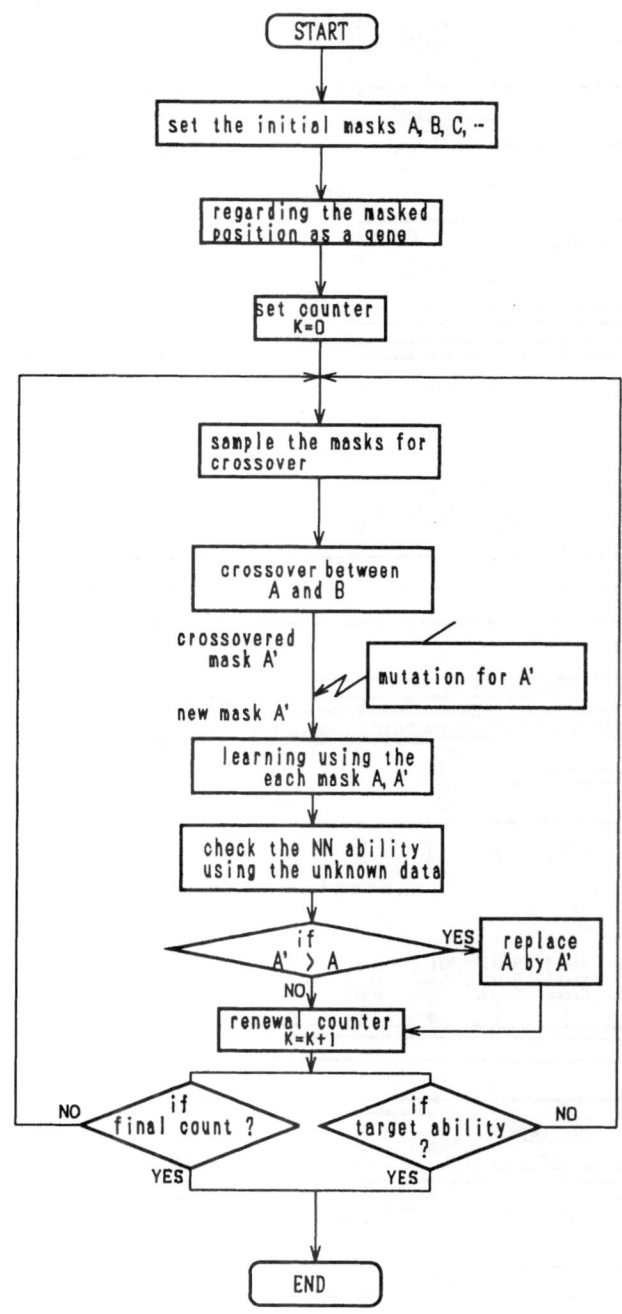

Fig.9. Flowchart of optimizing mask set by the GA.

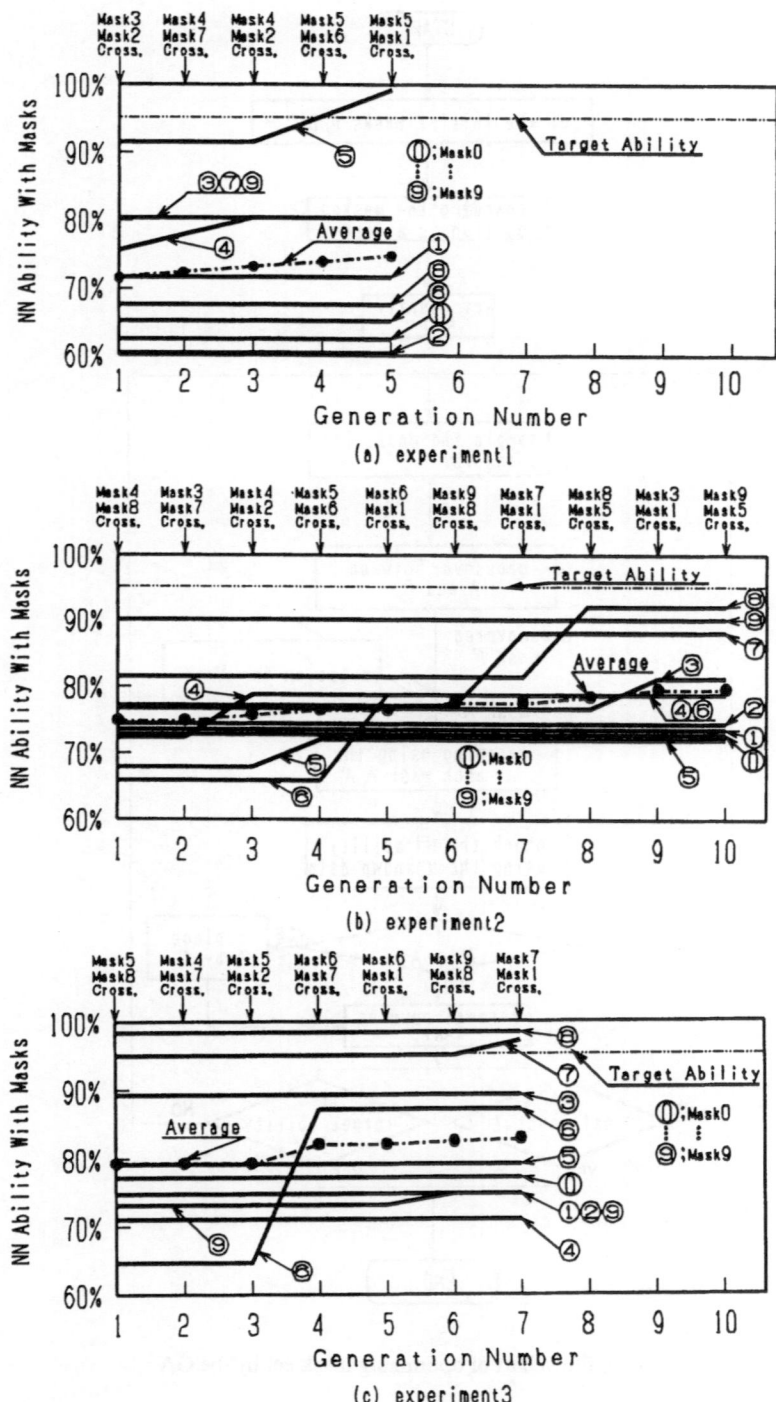

Fig.10. Transition of optimizing mask set by the GA.

every mask set, the every combination constructed by the least masked area can be considered as the mask set. So the inputs of the every mask set can be generated. Learning should be executed using each input. We have to investigate the generalization of the NN with the each mask set using unknown data. In this case, we decide the mask set which generates the inputs that show the highest generalization as the optimized one.

This procedure can be operated in the case of small number of combination. But the number of mask set can be calculated as 2^M (M is the number of the least masked area and here it is 16). When M increases, this procedure is no more effective to optimize the mask set. In this reason, proposed method using the GA is effective to systematize the neuro-money recognition method with mask sets and apply the NN to the commercial products. If the recognized object is replaced from US dollars to another kind of bill money, we can easily apply the neuro-bank notes recognition method with masks to them in a short period by the GA.

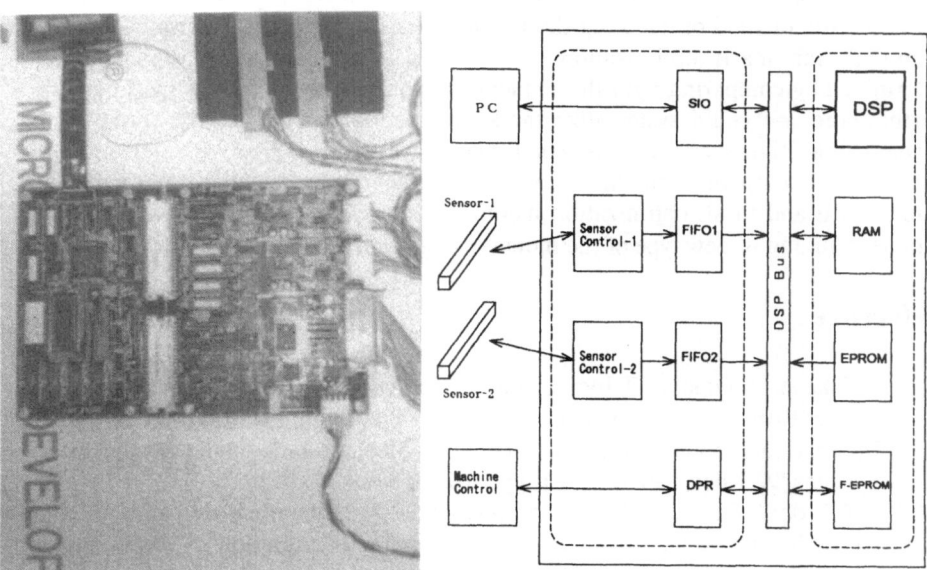

Fig.11. Neuro-recognition board.　　　　Fig.12. Block diagram of neuro-board

5. Neuro-Money Recognition Board Using DSP

We have considered the neuro-money recognition with optimized masks by the GA in the commercial products and developed the high-speed neuro-recognition board as shown in Fig.11. In this figure, left side is the DSP circuit and the right one is the interface circuit for the sensor. Still more, Fig.12 shows the block diagram of the neuro-recognition board. Neuro program boots up from EPROM. Neuron's weights are saved in flash memory and they can be renewed by extra-host computer.

Here, the adopted DSP has the exponential function which is used in the sigmoid function $(f(x)=1/(1+\exp(-x)))$ as a library. This enables us to make easy for implementation of neuro-algorithm from EWS or another large computer to the real time systems. Its transaction speed is more than 10 times compared with the conventional recognition machine. When whole transaction of the neuro-recognition method with masks is regard as 100, it has been shown that edge detecting needs 28, mask calculation needs 35, and neuro-calculation needs 37. This DSP runs under 33 MHz machine clock. Core parts of this neuro-recognition algorithm are written in assemble language.

6. Conclusions

In this paper, we have adopted the GA to the neuro-money recognition method with masks and have shown its effectiveness using US dollars. First, we have described basic idea of the neuro-money recognition method with masks. Then we discussed the method of adopting the GA to mask optimization. By use of this new technique for mask optimization, we showed that effective masks could be determined by comparing with the random masks in a short period ,experimentally. Furthermore, we could systematize mask optimization using the GA. Finally, we developed a high-speed neuro-recognition board with DSP. We confirmed its effectiveness compared with the conventional recognition machines. We hope this neuro-board and mask optimization method by the GA enable us to accelerate the commercializing of new type of the money-recognition machine.

Reference

[1] F.Takeda, S.Omatu, T.Inoue, and S.Onami, "High Speed Conveyed Bill Money Recognition with Neural Network", Proceedings of the IMACS/SCINE International Symposium on Robotics, Mechatronics and Manufacturing Systems '92 Kobe, Japan, Vol.1, pp.16-20, 1992.

[2] F.Takeda, S.Omatu, T.Inoue, and S.Onami, "A Structure Reduction of Neural Network with Random Masks and Bill Money Recognition", Proceedings of the 2nd International Conference on Fuzzy Logic and Neural Networks, IIZUKA, Japan, Vol.2,pp.809-813, 1992.

[3] F.Takeda and S.Omatu, "Bank Note Recognition System Using Neural Network with Random Masks ", Proceedings of the World Congress on Neural Networks, Portland, USA, Vol.1, pp.241-244, 1993.

[4] F.Takeda and S.Omatu, "Recognition System of US Dollars Using a Neural Network with Random Masks", Proceedings of the International Joint Conference on Neural Networks, Nagoya, Japan, Vol.2, pp.2033-2036, 1993.

[5] B.Widrow,R.G.Winter, and R.A.Baxter, "Layered Neural Nets for Pattern Recog- nition", IEEE Trans. on Acoust., Speech & Signal Process., Vol.36, No.7, pp. 1109-1118, 1988.

[6] D.E. Goldberg, Genetic Algorithms in Search, Optimization and Machine Learning, Addison-Wesley, New York, 1989.

[7]   F.Takeda, S.Omatu, S.Onami, T.Kadono, and K.Terada,"A Paper Currency Recognition Method by a Small Size Neural Network with Optimized Masks by GA", Proceedings of IEEE World Congress on Computational Intelligence, Orlando, USA, Vol.7, pp.4243-4246 1994.

[8]   F.Takeda, S.Omatu, S.Onami, T.Kadono, and K.Terada,"A Paper Currency Recognition Method by a Neural Network Using Maks and Mask Optimization by GA," Proceedings of World Wisemen /women Workshop On Fuzzy Logic And Neural Networks /Genetic Algorithms of IEEE / Nagoya University, Nagoya, Japan 1994.

[9]   Kitano,Genetic Algorithm, Sangyo Tosyo, pp.44-60, 1993 (in Japanese)

Genetic-Fuzzy Systems for Financial Decision Making

Suran Goonatilake

John A. Campbell

Department of Computer Science
University College London
Gower Street, London WC1E 6BT

Nesar Ahmad

Dept. of Electrical Engineering
Indian Institute of Technology
New Delhi 110016

Abstract : This paper describes the use of genetic algorithms for inducing fuzzy rule-bases within the context of decision support systems for financial trading. The genetic algorithm part of the procedure is based on Packard's algorithm for complex data analysis. The fuzzy pre-processing of the data is achieved by using a Single Linkage clustering algorithm in conjunction with an heuristic cluster selection mechanism. We believe that this hybrid approach has advantages over other 'black-box' machine learning procedures in that it produces *transparent* decision models that are easily understood by decision-makers. Further, the induced decision models lend themselves naturally to judgmental revisions by decision-makers.

1. Introduction

Intelligent systems are now being used to automate a wide variety of decision making tasks in business and finance. Neural networks have recently been applied successfully in areas such as currencies and derivatives forecasting, insurance underwriting and portfolio management [14], [21]. However, for some applications such as credit evaluation *explanations* of the reasoning process are of paramount importance, sometimes even being a legal requirement [22]. As neural networks have inherent difficulties in providing explanations, business users are increasingly examining approaches of the rule induction type which provide explicit explanations of reasoning.

The approach presented in this paper uses genetic algorithms to induce fuzzy decision rules operating on data with 'linguistic' categories such as *low, medium* and *high*. This we believe produces extremely easy to understand 'transparent' decision models which can be appreciated by technical personnel and high level strategic decision-makers alike. Further, the induced decision models naturally lend themselves to judgmental revisions by decision-makers. The genetic algorithm part of the procedure is based on Packard's algorithm for complex data analysis [10]. The fuzzy pre-processing of the data is achieved by using a Single Linkage clustering (SLINK) algorithm in conjunction with an heuristic cluster selection mechanism.

Although we discuss this approach with an example from the area of decision support in financial trading, this method evidently has wide applications in other areas of business decision making including credit evaluation, corporate risk assessment and insurance underwriting.

We now discuss properties that are required from an ideal decision support system for financial trading.

1.1 Learning from the market

As financial traders know, the characteristics that define a particular market can change significantly over a relatively short time period [15]. For example, for a few months, rises in interest rates may strengthen a currency, while it is possible that in following months, rises in interest rates may actually weaken a currency [16]. Market practitioners are well aware of such character changes in markets and can quickly adjust to these. Because of these changes of character in financial systems, a trading decision support system should, ideally, have the ability to adapt to such changes and be able to make successful trading recommendations before and after such changes. In order for this to happen, the trading decision support system should have the capability to learn continuously from the market, i.e. to induce new relationships constantly.

1.2 Robustness

Financial market data are typically very noisy [18]. A good financial decision support system should be robust enough to find patterns in such data. In financial systems it is rare that a particular set of conditions (say, particular values of market indicators), if observed again, will produce the same future market behaviour. Because of this non-deterministic nature of financial markets, strict deterministic rules will fail to perform well in the domain. For example, if by observing past data one induces a rule that if the Pound/Dollar exchange rate is 1.546 and the volume of contracts traded is 23898 then the market will rise, it is probable that because of these very specific conditions the future market data will never match these conditions exactly. It is therefore vital that a good trading decision system should have the capability to induce patterns in a fuzzy or statistical manner.

1.3 Explanation

When financial trading decisions involve large amounts of money, reassurance as to the soundness of the decision-making procedure is needed. The ability to cite the exact conditions and reasoning of a trading decision is often required by senior managers in financial organisations. As computer-assisted decision making is still relatively uncommon, most managers do not entirely 'trust' machine-generated trading decisions and require an understanding of the decision process in a format that they can understand. Because of this factor, some investment managers remain suspicious of the use of neural networks and other 'black box' techniques for making trading decisions [1].

Our approach attempts to satisfy all of the above considerations (learning, robustness, explanation). Sections 2 and 3 describe the genetic algorithm, Section 4 describes the fuzzy pre-processing method, Section 5 describes the induction of fuzzy rule-bases and Section 6 describes the application of the technique in financial trading.

2. Packard's System for Complex Data Analysis

Packard's genetic algorithm [10] can be viewed as a *model searching mechanism* which searches a very large space of possible models to find a good set of models that can capture underlying regularities of the given system being studied.

There are three main components in Packard's modelling system.

1. A code or representation scheme for the models
2. A mechanism to evaluate the usefulness (fitness) of the models
3. A mechanism to generate new models

2.1 Data

Let us assume the data to be a collection of pairs $(\bar{x}, y)$, $((x_1 \ldots x_n), y) = (\bar{x}, y)$ where each $\bar{x}$ is a set of independent variables (*features*) and where y is the corresponding dependent variable (*classification variable*). Both the independent and dependent variables have to have discrete states, and if the source is continuous the values have to be discretised or 'binned'.

The aim of the algorithm is to search for states of the independent variables, $\bar{x}$, which on the average have a high correlation with particular desired states of y, the dependent variable. The induced patterns will take the form of a set of hypotheses or models, each of the form, *"when some subset of the independent variables satisfies particular conditions, a certain behaviour of the dependent variable is to be expected."* In a market forecasting context, the dependent variable will typically be a future (discretised) state of the system such as 'the market in 10 days (BUY or

SELL)' and the independent variables will be (discretised) states of technical trading indicators such as 'Open interest *low*' and 'Volume *high*' etc.

When the algorithm is used in a forecasting context, Packard describes the algorithm as 'searching for pockets of predictability'. It is assumed that the majority of the search space is in fact non-predictable, and the algorithm attempts to fit models to the predictable parts (pockets) of the search space. This non-predictability can be due to external (measurement) noise, dynamical noise (chaotic behaviour), or lack of sufficient data [10].

In the trading context, one can use the model to discover BUY, SELL trading decisions depending on the states of technical trading indicators.

2.2 Representation of Models

The representation of models or *conditional sets* in Packard's system is in the familiar *disjunctive normal form* [12], which specifies relationships between entities in terms of AND, OR relations. A conditional set or model contains as many 'condition positions' as there are independent co-ordinates, n, identifying each of them with one of the co-ordinates. Each 'condition position' will be allowed to take on either a value of *, indicating no condition is set for the corresponding co-ordinate, or a sequence of numbers $(c_1, \ldots, c_k)$ indicating OR'ed values of the corresponding co-ordinate. For example,

$$(*, (5,9), *, *, 7, *, *) \sim X_c$$

indicates that the conditional set X_c will be true if the second co-ordinate has a value of either 5 or 9, and the fifth co-ordinate has a value of 7. It will ignore the values of the other co-ordinates.

2.3 Searching for Good Decision Models

If the aim of the algorithm is to find *good models* or conditional sets, then there must be a mechanism for evaluating the *goodness* or 'fitness' of a given model. This means to find the level of correlation between the states of the independent variables and the *target* dependent variable.

Let N_c be the total number of points in the conditional set X_c (the set of points that satisfy all the specified conditions). We then construct our empirical estimate of the conditional probability distribution of y values given the values of $\vec{x} \in X_c$, where N_c is the number of points in X_c. $\delta(y - y')$ is 1 if $y = y'$ and 0 otherwise.

$$P_c(y) = \frac{1}{N_c} \sum_{(x,y) \in X_c} \delta(y - y')$$

Packard [10] has also introduced a 'devaluing' operator to guard against the building of conditional sets that have very small numbers of data points in them, and hence to reduce the effects of statistical flukes. A term proportional to $\frac{1}{N_c}$ is introduced here to achieve this devaluation.

The fitness F_c of a model or conditional set with the devaluation operation is therefore defined as

$$F_c(y) = P_c(y) - \frac{\alpha}{N_c}$$

where α is a parameter to adjust the dependence on N_c.

2.4 The Genetic Algorithm

The mechanism to generate new conditional sets or models in Packard's system is via a standard kind of genetic algorithm. The genetic cycle is:

1. Initialise the population with a random set of conditional sets.

2. Calculate the fitness of each conditional set via the fitness evaluation procedure described above.

3. Discard a fraction of the population with low fitness, and replace the deleted members with alterations of the remaining population, using the genetic operators.

4. Repeat 2 and 3 until good conditional sets are found.

The mutation operators Packard uses are:

1. Picking a new co-ordinate $* \rightarrow (a1, a2)$

$$(*,(5,9),*,*,7,*,*) \rightarrow (*,(5,9),*,*,7,*,3)$$

2. Deleting a co-ordinate

$$(*,(5,9),*,*,7,*,*) \rightarrow (*,(5,9),*,*,*,*,*)$$

3. Changing a co-ordinate

$$(*,(5,9),*,*,7,*,*) \rightarrow (*,(5,9),*,*,2,*,*)$$

Crossover operations are performed preserving the positions of the conditional sets. For example, the two conditional sets C1 and C2 may produce the new conditional sets C3 and C4.

$$C_1(*,(5,9),*,*,7,*,*) \qquad C_3(*,(2,3),*,3,*,*,*)$$

$$\rightarrow$$

$$C_2(*,(2,3),*,3,*,*,4) \qquad C_4(*,(5,9),*,*,7,*,4)$$

3. Implementation of Packard's System

Although Packard has not referred to his system as a Classifier System, architecturally it is very similar to a Classifier System. If one replaces the term 'conditional codes' with *production rules*, one essentially gets a classifier system. A departure from the traditional classifier system [7] is the lack of chains of inferences and the associated credit assignment mechanisms [7]. In contrast, Packard's system evaluates each rule individually, thus shifting the emphasis from finding a '*set* of good classifiers' to finding 'good *individual* classifiers'.

There are two modes of operation in our adaptation of Packard's system: *symbolic* and *fuzzy*. In the symbolic mode, the genetic algorithm operates on symbolic (crisp) representations of data and induces *individual* decision rules. In the fuzzy mode, the genetic algorithm operates on fuzzified data and induces *collections* of fuzzy rules (rule-bases). The symbolic pre-processing of data and the subsequent fuzzification of data is achieved using a SLINK clustering algorithm (described in Section 4). Both modes use the same basic representation of rules and rule-modification operators.

3.1 Rule Representation

An example of our representation is as follows, where our dependent variable (target variable) is *risk* with two possible decision states: high or low.

For example if we choose three independent variables to be used for the genetic search process — *max-speed, age-of-car* and *age-of-driver. max-speed* has three possible states, [low medium high], *age-of-car* has two possible states, [old new] and *age-of-driver* has three possible states, [young middle senior].

An example of a representation of a production rule is,

```
[ IF [max-speed [high] AND [age-of-car [new]] AND
[age-of-driver [young]  THEN [risk [high] ]
```

If a rule has a term which has no expression within it, e.g. max-speed [ ], then that term is not evaluated.

3.2 The Genetic Algorithm

Our implementation of the genetic algorithm cycle has the following seven standard steps.

1. Initialisation of a population of (random) Rules.

2. Evaluation of fitness of each Rule in the population.

3. Selection of parent Rules for alteration.

4. Creation of new Rules by Crossover and Mutation operators.

5. Deletion of the old rule population.

6. Creation of a new population by inserting altered rules and the fittest rules.

7. Go to 3 until a satisfactory rule(s) is found or a specified number of iterations have been completed.

3.2.1 Rule Initialisation

The variables for each rule in the population (each represented as a list) are initialised with randomly-chosen values which are looked up in a reference table (feature-list) which stores each variable's permissible states. For example for the variable max-speed, the reference table (feature-list) is looked up and from the three permissible states [low medium high], one state is randomly chosen.

3.2.2 Rule Evaluation

Each rule in the population is evaluated using Packard's fitness evaluation function $F_c(y)$ described in Section 2.3.

At this stage of rule evaluation we also select a portion of the *fittest members* from the current population for inclusion in the next new population.

3.2.3 Parent Selection

Once the fitness for each rule has been established, we use the roulette-wheel selection procedure [4] to select the best 'parent' rules for alteration. We have not experimented with any other parent selection schemes, and this particular scheme was chosen because of its proven effectiveness in solving a variety of real-world problems [4].

3.2.4 Rule Alteration: crossover and mutation

The rules that are returned by the roulette-wheel procedure are then altered via the crossover and mutation operators to create new rules.

The crossover operation swaps the conditions of rules with other conditions of other rules, at the same conditional locations. The crossover rate (C_r) which is set by the user determines the probability of a crossover operation occurring at a particular conditional point.

If there are two population members (chromosomes) c1, c2,

```
c1 : IF [max-speed [high]] AND [age-of-car [new]] THEN
[risk   [high] ]

c2:  IF [max-speed [low]] AND [age-of-car [old]]  THEN
[risk low ]]
```

the effect of crossover can result in forming the following two new rules C3 and C4;

```
c3:  IF [max-speed [low]] AND [age-of-car [new]] THEN
[risk [low]]

c4:  IF [max-speed [high]] AND [age-of-car [old]] THEN
[risk high]]
```

There are three mutation operators in the system. The probability of a mutation operation being applied is determined by the user-specified mutation rate *M*. The three mutation operators are:

1. Picking a new co-ordinate

```
age-of-driver []  →  age-of-driver [young]
```

2. Deleting a co-ordinate

```
age-of-driver  [young]  →  age-of-driver []
```

3. Changing the value of a co-ordinate

```
age-of-driver [young]  →  age-of-driver [senior]
```

4. Fuzzy Data Pre-processing

Most decision-makers in finance and business commonly use linguistic categories (e.g. low, high, large) to describe complex relationships in their domain. We therefore ideally need a mechanism to convert 'raw' data from a domain (e.g. price, volume and open interest data) into such linguistic symbolic descriptions. We use a relatively simple method based on the use of a clustering algorithm to convert such data into linguistic descriptions. It is on these linguistic descriptions that the genetic algorithm will operate on.

The starting point for the pre-processing method is for the user to specify linguistic 'labels'. These labels are for the symbolic categories into which the algorithm will subsequently classify raw data. Examples of these labels are low, medium, high and small, moderate and big. The linguistic categories should be specified in an increasing order e.g.: low , medium, high.

Once the order of the labels are specified, a clustering algorithm is applied to the raw market data. The clustering algorithm used is the Single Linkage Clustering Method (SLINK). The SLINK clustering algorithm is in the family of *nearest neighbour* techniques, which iteratively grows data points that are nearest to each other into clusters. We chose the SLINK algorithm because it is a computationally efficient clustering procedure [6], especially compared with neural network clustering methods [9], [2]. A public domain implementation of the SLINK algorithm written by Stolcke [11] is used for all clustering operations.

4.1 The SLINK Clustering Algorithm

The following is a description of the Single Linkage Clustering (SLINK) algorithm [6] that is used for clustering the data.

The M objects will be arranged in order so that each cluster is a contiguous sequence of objects. The Ith object in this new order will be denoted by $O(I)$ and distance is denoted by D. A *gap* $G(I)$ is associated with the Ith object in the order. These gaps determine the boundaries of the clusters.

Step 1. Let $O(1)$ be any object. Let $G(1) = \infty$.

Step 2. Let $O(2)$ be the object closest to $O(1)$. Let $G(2)$ be the distance between $O(2)$ and $O(1)$.

Step 3. For each $I(3 \leq I \leq M)$ let $O(I)$ be the object, not among $O(1), O(2), \ldots, O(I-1)$. That is, for some $K(1 \leq K \leq I - 1)$

$$D[O(I), O(K)] \leq D(J, L),$$

where J ranges over $O(1), \ldots, O(I-1)$ and L ranges over the remaining objects. The gap $G(I)$ is set equal to this minimum distance $D[O(I), O(K)]$.

Step 4. The cluster $O(L1) - O(L2)$, containing objects $O(L1)$, $O(L1+1), \ldots, O(L2 - 1)$, $O(L2)$ is associated with gap $G(I)$, where $(L1, L2)$ is the maximal interval including I, such that $G(J) \leq G(I)$ for all J with $L1 < J \leq L2$.

4.2 Cluster Selection

The clustering algorithm returns a tree structure of the clustered items as a list of lists. The next task is to select clusters that correspond to the linguistic items. The cluster selection algorithm operates using two main heuristics.

1. The clusters that define the linguistic categories will have a larger number of data points than clusters that do not correspond to the linguistic categories.

2. The clusters corresponding to the linguistic categories will roughly divide the total number of data points among them.

The algorithm for selecting the clusters is as follows:

1. Order the linguistic items in terms of increasing value (e.g. low medium high).

2. Create a *slots list* (a list of lists) where the number of elements equals the number of linguistic categories and initialise these lists with zero elements.

3. Calculate the 'ideal' cluster length. This is computed based on the heuristic that the clusters corresponding to the linguistic categories equally divide all the data points.

4. ideal cluster length = total number of data points / number of linguistic categories

Start from the root of the cluster tree and compare the slot lists with the clusters,

IF the cluster has common elements with any of the slots in the slots list, AND the cluster is closer to the ideal cluster length, THEN replace the contents of the slot with the current cluster

ELSE

IF there are no common slots THEN,

(a) Find the worst slot from the slots list (the worst slot is the slot that has a maximal length difference from the ideal cluster length)

(b) IF the current cluster is closer to the ideal cluster than the worst slot, THEN replace the contents of that slot with the current cluster

Repeat 4 until the leaves of the cluster tree are reached.

4.3 Defining the Fuzzy Sets

The above described cluster selection algorithm selects clusters that correspond to user-specified linguistic categories. Taking the numeric ranges of the selected clusters, one can then classify new datapoints into different linguistic categories. However, these cluster boundaries are 'crisp' where linguistic categories have sharp, abrupt divisions. In the *symbolic mode* of our system, rules are induced using these 'crisp' symbolic descriptions.

In the *fuzzy mode*, the cluster boundaries are smoothed to produce fuzzy descriptions. We achieve this by defining triangular fuzzy membership functions using the cluster ranges as 'anchor points' (described in detail below).

We will illustrate our fuzzy membership definition process starting with cluster ranges obtained from the cluster selection algorithm corresponding to low, medium and high derived from volume (V) data (see figure 1).

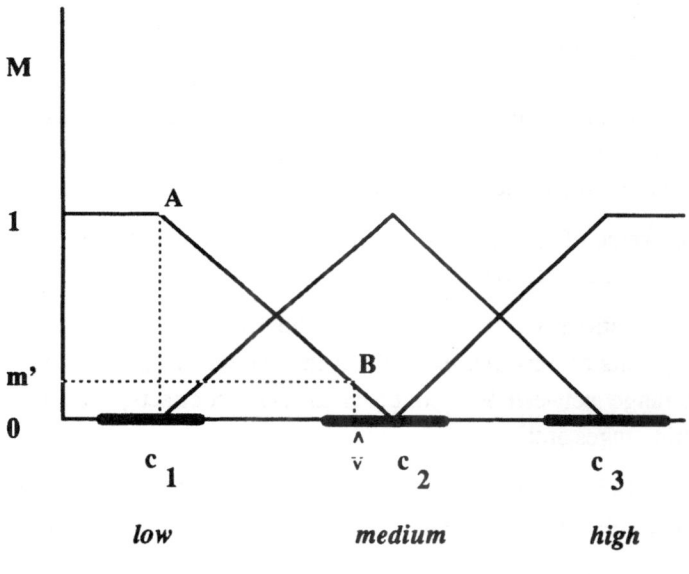

Figure 1

Let the fuzzy sets of volume be $v_1 = low$, $v_2 = medium$, $v_3 = high$. We define the range of the fuzzy membership functions to be between 0 and 1.

Let c_1 be the midpoint of the first set v_1 (low), let c_2 be the midpoint of the middle set v_2 (medium), and c_3 the midpoint of the set v_3 (high).

We use the following algorithm for defining the shapes of the fuzzy sets:

If it is the first set, it has a trapezoidal shape where the line opposite the base and the line opposite the right angle take values corresponding to fuzzy memberships. The line opposite the base has a membership $\mu = 1$ where $v \leq c_1$.

The line opposite the right angle joins the base at the mid point of the next set (c_2). The membership in this section $\mu = (c_2 - \hat{v})/ (c_2 - c_1)$.

If the set is a middle set then the membership will be defined as a triangle where the vertex has a membership $\mu=1$ at a point perpendicular to the midpoint. The two sides of the triangle join the midpoints of the two adjacent fuzzy sets. The upward slant is calculated by $\mu = (\hat{v} - c_1)/(c_2 - c_1)$. The downward slant is calculated by $\mu = (c_3 - \hat{v})/(c_3 - c_2)$.

If the set is the last set, then the membership has a trapezoidal shape where the line opposite the right angle joins the base at the midpoint of the previous set and joins the line perpendicular to the base at the mid-point of the set. The membership in this section is $\mu = (\hat{v} - c_2)/(c_3 - c_2)$. The line opposite the base has a membership $\mu = 1$ where $\hat{v} >= c_3$.

4.4 The Universe of Discourse and Fuzzification

In our method, the minimum value of the universe of discourse corresponds to the minimum value in the cluster with the lowest values and the maximum corresponds to the maximum value in the highest cluster.

We discretise the universe of discourse (UoD) into equal segments to simplify the computations involved in fuzzy reasoning.

In the following applications we have discretised the UoD into an arbitrary number of (13) equal segments or bins. For example, if the variable volume, v, has chosen clusters with a range between 911 contracts to 4491 contracts, then the thirteen corresponding bin ranges are;

[911.0--1186 1186--1461 1461--1737 1737--2012 2012--2287 2287--2563 2563--2838 2838--3114 3314--3389 3389--3664 3664--3940 3940--4215 4215--4491]

Referring to the fuzzy set definition in figure 1, and the above universe of discourse, a volume value of 912 will have a membership of 1 for the *low* fuzzy set, and a membership of 0 for the fuzzy sets *medium* and *high*. Similarly a volume value of 4490 will have a membership of 1 for the *high* fuzzy set. An example of definitions of the degree of memberships for all three fuzzy sets for the above discretised thirteen ranges are:

[low   [1.0 1.0 1.0 0.8 0.6 0.5 0.3 0.1 0.0 0.0 0.0 0.0 0.0]]

[medium [0.0 0.0 0.0 0.1 0.3 0.4 0.6 0.8 0.9 0.6 0.4 0.1 0.0]]

[high   [0.0 0.0 0.0 0.0 0.0 0.0 0.0 0.0 0.0 0.3 0.5 0.8 1.0]]

Once the definitions for the fuzzy sets are obtained, classifying a new raw data item into fuzzy values is straightforward. Classifying a new value involves looking up the corresponding range (bin) in the discretised universe of discourse and then looking up the corresponding fuzzy values from the defined fuzzy sets. For example taking the above Volume data and fuzzy definitions, suppose that we need to find the fuzzy values for the data item of 2000 contracts. First the corresponding bin is found in the UoD - the bin range is the 1737-2012 bin (the fourth bin). Then the fuzzy values are obtained by looking up fuzzy sets corresponding to this bin range. These fuzzy values are: 0.8 low, 0.1 medium and 0.0 high.

4.5 Defining Membership Functions for the Rule Consequents (decisions)

The membership functions for the consequents, the decisions, (buy or sell) are defined heuristically. The trading decisions are defined as having a range of [-3,+3] where the negative values indicate a SELL decision while the positive values indicate a BUY decision (see figure 2). A membership function, DO-NOTHING, reflecting the decision not to trade has also been defined. The numerical values indicate the level of confidence of the decision (e.g. - 2.9 a definite SELL decision, - 0.8 a less definite SELL decision).

The Membership function for Trading Decision (TD) is TD_i:

UoD	[-3	-2.0	-1.0	0.0	+1.0	+2.	+3]
TD_1 SELL	[1	0.8	0.2	0.0	1.0	0.0	0]
TD_2 DO-NOTHING	[0	0.2	0.8	1.0	0.8	0.2	0]
TD_3 BUY	[0	0.0	0.0	0.0	0.2	0.8	1]

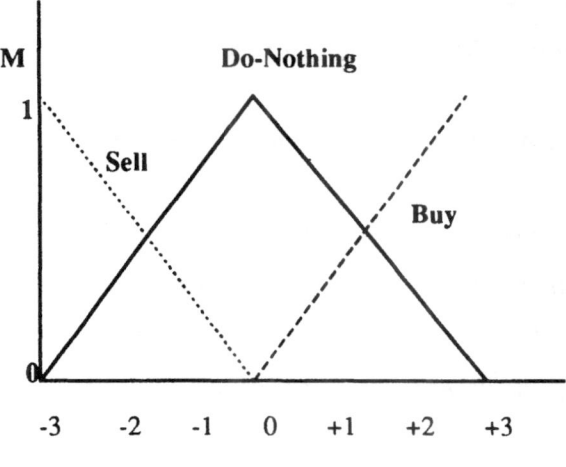

Figure 2

5. Inducing Fuzzy Rule-Bases

There is a very large number of possible fuzzy models that can be constructed using a given set of variables. We use the genetic algorithm as a mechanism to search through the very large space of possible fuzzy models (rule bases) to find a set of good fuzzy models.

The genetic algorithm cycle in the fuzzy mode operates as follows. The population is firstly initialised with a random collection of fuzzy rule bases. At each iteration, each fuzzy rule base will make fuzzy inferences on fuzzified data. The final results are then passed through a threshold and the final trading decisions are obtained. These decisions are then compared with the known 'best' decisions using the past trading data, and the fitness of rule bases are calculated accordingly. The fuzzy rule bases are ranked in terms of their fitness, and afterwards mutation and crossover operations are performed to produce new rule bases.

Over time this procedure produces a collection of highly effective fuzzy rule bases.

It is the *aggregation* effect of *all fuzzy rules* (all of which may partially match the data) that results in a fuzzy system's ability to deal with brittleness [24]. Therefore, unlike in the case of the previous symbolic mode where we use the GA to find good *individual* rules, we now use the GA to find good *collections of rules* (fuzzy rule bases).

The genetic algorithm representation is different in this mode in that it now has a population of *rule bases* as opposed to a population of rules. The aim of the algorithm is to find good rule bases consisting of good predictive rules.

An example of two members of the population, GF1 and GF2 (each rule base has 4 rules) :

[**GF1** [ma-diff-1-20-fuzzy [positive]] AND [oi-rsi-14-fuzzy [high]] THEN [action [BUY]]

[ma-diff-1-20-fuzzy [negative]] AND [oi-rsi-14-fuzzy [low]] THEN [action [SELL]]

[ma-diff-1-20-fuzzy [positive]] AND [oi-rsi-14-fuzzy [medium]] THEN [action [BUY]]

[ma-diff-1-20-fuzzy [positive]] AND [oi-rsi-14-fuzzy [high]] THEN [action [BUY]]]

[**GF2** [ma-diff-1-20-fuzzy [positive]] AND [oi-rsi-14-fuzzy [high]] THEN [action [SELL]]

[ma-diff-1-20-fuzzy [negative]] AND [oi-rsi-14-fuzzy [high]] THEN [action [SELL]] [ma-diff-1-20-fuzzy [neutral]] AND [oi-rsi-14-fuzzy [low]] THEN [action [BUY]]

[ma-diff-1-20-fuzzy [neutral]] AND [oi-rsi-14-fuzzy [high]] THEN [action [BUY]]]

The evaluation of each member (rule-base) is performed as follows. We firstly apply Mamdani and Assilian's compositional rule of inference [23]. Then we use the *centre of area* method [24] as the defuzzification procedure to obtain the final result.

As defined by the universe of discourse of trading decisions, this result will have a range [-3,+3] (values close to -3 indicate a SELL decision, values closer to +3 indicate a BUY decision). We then apply a threshold infer the final decision (values <-2.2 SELL, values > +2.2 BUY).

After this, the Packard's fitness evaluation procedure (as detailed in section 2.3) is applied and the fitnesses of the rule bases are computed. That is, for each fuzzy rule base the fitness of its decisions are calculated where N_c are data entries that return values beyond the specified fuzzy threshold (values <-2.2 SELL, values > +2.2 BUY).

In the fuzzy mode, the crossover and mutation operators are modified to take account of the different structure of the population members. Crossover can occur both within an individual rule base as well as with members that are in a different rule base. The mutation operation is unchanged with the difference that the mutation probability rate is now applied to a rule base rather than an individual rule.

6. The Financial Trading Application

We now investigate the application of the above approach as a mechanism for supporting decisions in the domain of currency trading. The method is used to induce fuzzy rule-bases that operate on technical trading [18] indicators. Technical trading rules are used by a large number of traders and our method provides an automated method for discovering such trading knowledge. We do not envisage that this type of approach will completely *automate* the trading process, but instead take the view that it is a good *decision support* tool for traders. Traders may overrule or change the conclusions of the rules due to considerations external to the models (e.g. political events). Therefore, ideally, rules discovered by this approach will be firstly presented to a trader for judgmental revisions before being used for trading.

The moving average method is a commonly-used technical trading indicator. Generally two moving averages are used - a long period (e.g. the moving average of the last 200 day's prices) and a short period (e.g. the moving average of the last 10 day's prices), this is typically written as a 10-200 system. A 1-10 system would be one in which the long moving average was for a 10 day period and the short was the actual daily price. An example of a moving average system is given in figure 3. The general idea behind computing the moving averages is that they smooth the generally volatile time series, and provide an indication of the general trend of the market [18].

The moving average of prices is given by

$$MA_t = \frac{1}{N} \sum_{i=0}^{N-1} P_{t-i}$$

where N is the number of days, P is the price and MA_t is the moving average on day t. There are several ways [19], [18] of using moving averages for making trading decisions. One type of trading strategy is to execute BUY trades when the short moving average is higher than the long moving average, and to execute SELL trades when the short moving average is lower than the long moving average. The rules corresponding to these hypotheses are:

(E1) If the **short moving average** is higher than the **long moving average** then the market is likely to **rise** (action : BUY)

(E2) If the **short moving average** is lower than the **long moving average** then the market is likely to **fall** (action : SELL)

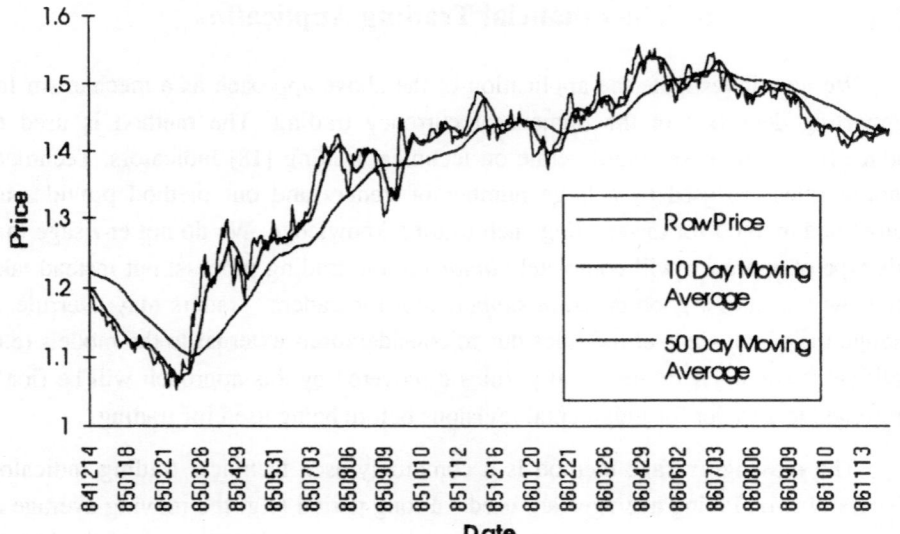

Figure 3

A variation of this approach is to execute trades when the moving averages cross each other. With this strategy a BUY trade is executed at the point when the short moving average becomes higher than the long moving average, and a SELL trade is executed at the point when the short moving average becomes lower than the long moving average.

The rules corresponding to these hypotheses are:

(E3) If the **short moving average** crosses the **long moving average** from *below* then the market is likely to **rise** (action: BUY)

(E4) If the **short moving average** crosses the **long moving average** from *above* then the market is likely to **fall** (action : SELL)

The effect of these rules can be seen in Figure 4. All the above moving average schemes are essentially based on measures reflecting the *difference* between the values of the two moving averages.

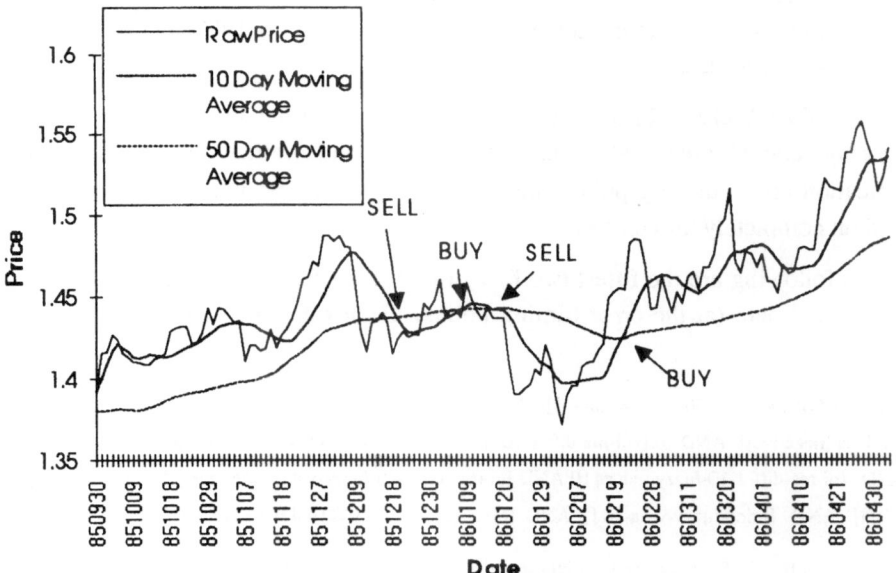

Figure 4: Trading Signals from Moving Average Model

We have used a simple measure of this difference,

$$MA_{diff} = \frac{SMA - LMA}{SMA}$$

where SMA is the short moving average, LMA is the long moving average and MA_{diff} is the measure of difference between the two moving averages.

A possible trading strategy based on this difference measure is,

If the MA_{diff} is **positive** then the price is likely to **rise** (action : BUY)

If the MA_{diff} is **negative** then the price is likely to **fall** (action : SELL)

The aim of this work is to use the genetic algorithm to discover fuzzy trading rules similar to ones described above. The data used are prices of the British Pound against the US dollar. The independent variables are moving-average differences of the closing prices, volume, and open interest and the dependent variable is the price after 10 days (UP or DOWN). Data from 1982 to 1984 are used to derive the cluster ranges while data from 1984 to 1987 are used to induce the rules. Data from 1987 to 1988 are used as a validation set while data from 1988 to 1992 are completely unseen data used for testing.

We follow Weiss [12] and attempt to avoid overfitting the training data by monitoring classification rates on the training and validation sets. At the classification error 'turning points' the fittest fuzzy rule base is selected, and then applied on completely unseen data.

The following are the fittest two fuzzy rule bases (FR1,FR2) induced using the British Pound data (at the error turning points). Each rule-base consists of 4 fuzzy rules.

[[FR1[G1[ma-diff-1-50-fuzzy-values []] AND [ma-diff-1-100-fuzzy-values[]] AND [ma-diff-1-200-fuzzy-values [negative]] AND [vol-ma-diff-1-10-fuzzy-values []] AND [vol-ma-diff-1-20-fuzzy-values []] AND [oi-ma-diff-1-10-fuzzy-values []] AND [oi-ma-diff-1-20-fuzzy-values [neutral]] AND [price-fuzzy-vola-20 []] AND [price-fuzzy-vola-50 []] AND [price-fuzzy-vola-100 []] THEN [action [SELL]]]

[G2[ma-diff-1-50-fuzzy-values []] AND [ma-diff-1-100-fuzzy-values []] AND [ma-diff-1-200-fuzzy-values [negative]] AND [vol-ma-diff-1-10-fuzzy-values []] AND [vol-ma-diff-1-20-fuzzy-values []] AND [oi-ma-diff-1-10-fuzzy-values []] AND [oi-ma-diff-1-20-fuzzy-values [neutral]] AND [price-fuzzy-vola-20 []] AND [price-fuzzy-vola-50 []] AND [price-fuzzy-vola-100 []] THEN [action [SELL]]]

[G3[ma-diff-1-50-fuzzy-values []] AND [ma-diff-1-100-fuzzy-values []] AND [ma-diff-1-200-fuzzy-values [negative]] AND [vol-ma-diff-1-10-fuzzy-values []] AND [vol-ma-diff-1-20-fuzzy-values []] AND [oi-ma-diff-1-10-fuzzy-values []] AND [oi-ma-diff-1-20-fuzzy-values [negative]] AND [price-fuzzy-vola-20 []] AND [price-fuzzy-vola-50 []] AND [price-fuzzy-vola-100 []] THEN [action [SELL]]]

[G4[ma-diff-1-50-fuzzy-values []] AND [ma-diff-1-100-fuzzy-values []] AND [ma-diff-1-200-fuzzy-values [negative]] AND [vol-ma-diff-1-10-fuzzy-values []] AND [vol-ma-diff-1-20-fuzzy-values []] AND [oi-ma-diff-1-10-fuzzy-values []] AND [oi-ma-diff-1-20-fuzzy-values [negative]] AND [price-fuzzy-vola-20 []] AND [price-fuzzy-vola-50 []] AND [price-fuzzy-vola-100 []] THEN [action [SELL]]]

[FR2[G1[ma-diff-1-50-fuzzy-values []] AND [ma-diff-1-100-fuzzy-values []] AND [ma-diff-1-200-fuzzy-values [negative]] AND [vol-ma-diff-1-10-fuzzy-values []] AND [vol-ma-diff-1-20-fuzzy-values []] AND [oi-ma-diff-1-10-fuzzy-values []] AND [oi-ma-diff-1-20-fuzzy-values [neutral]] AND [price-fuzzy-vola-20 []] AND [price-fuzzy-vola-50 []] AND [price-fuzzy-vola-100 []] THEN [action [SELL]]]

[G2[ma-diff-1-50-fuzzy-values []] AND [ma-diff-1-100-fuzzy-values []] AND [ma-diff-1-200-fuzzy-values [negative]] AND [vol-ma-diff-1-10-fuzzy-values []] AND [vol-ma-diff-1-20-fuzzy-values []] AND [oi-ma-diff-1-10-fuzzy-values []] AND [oi-ma-diff-1-20-fuzzy-values [neutral]] AND [price-fuzzy-vola-20 []] AND [price-fuzzy-vola-50 []] AND [price-fuzzy-vola-100 []] THEN [action [SELL]]]

[G3[ma-diff-1-50-fuzzy-values []] AND [ma-diff-1-100-fuzzy-values []] AND [ma-diff-1-200-fuzzy-values [negative]] AND [vol-ma-diff-1-10-fuzzy-values []] AND [vol-ma-diff-1-20-fuzzy-values []] AND [oi-ma-diff-1-10-fuzzy-values []] AND [oi-ma-diff-1-20-fuzzy-values [negative]] AND [price-fuzzy-vola-20 []] AND [price-fuzzy-vola-50 []] AND [price-fuzzy-vola-100 []] THEN [action [SELL]]]

[G4[ma-diff-1-50-fuzzy-values []] AND [ma-diff-1-100-fuzzy-values []] AND [ma-diff-1-200-fuzzy-values [negative]] AND [vol-ma-diff-1-10-fuzzy-values []] AND [vol-ma-diff-1-20-fuzzy-values []] AND [oi-ma-diff-1-10-fuzzy-values []] AND [oi-ma-diff-1-20-fuzzy-values [negative]] AND [price-fuzzy-vola-20 []] AND [price-fuzzy-vola-50 []] AND [price-fuzzy-vola-100 [high]] THEN [action [SELL]]] 2 0.635294]]

The system in the fuzzy mode produced 61% correct trades and 53% correct trades in the symbolic mode. We also undertook a detailed study of assessing human trader performance in the same foreign exchange prediction task (details of which are beyond the scope of this paper). The decisions made by the human trader were correct 64.2 % of the time.

Ideally the fuzzy rule-bases generated by the genetic algorithm should be revised judgmentally by a domain expert. As the model does not contain information external to it (e.g. political events), a trader may examine the rules and change any conditions if he or she wishes to.

Additional enhancements may include the selection of trading rules based on a more comprehensive fitness function. At present a fuzzy rule-base is deemed *fit* if it exhibits two behavioural attributes; firstly it must have high percentages of correct classifications and secondly there must be a sufficient number of past patterns to justify the particular rule-bases. Further extensions can include adjustments to the fitness function based on a term proportional to how much profit each rule-base makes. Such an addition may improve the results as it is claimed by traders that a vast of majority of their profits are actually generated by a very small number of trades.

Although we have demonstrated this approach as a mechanism for discovering knowledge in financial trading, it evidently has many applications in several other domains where explicit and easy to understand explanations of decision models is a prime concern. We are currently evaluating this approach in the areas of credit evaluation and insurance risk assessment.

References

[1]   Chartists tame chaos. The Economist, Aug 15 1992.

[2]   R. Beale and T. Jackson. *Neural Computing – an Introduction*. Adam Hilger, Bristol, 1990.

[3]   D. Bunn and G. Wright. Interaction of judgmental and statistical forecasting methods: Issues and analysis. *Management Science*, 37(5), 1991.

[4]   L. Davis. *Handbook of Genetic Algorithims*. Van Nostrand Reinhold, New York, 1991.

[5]   K.F. Wallis et al. *Models of the UK Economy: Reviews 1-5*. Oxford University Press, Oxford, 1988.

[6]   J. Hartigan. *Clustering Algorithms*. John Wiley and Sons, New York, 1975.

[7]   J. Holland. Genetic algorithms and classifier systems: Foundations and future directions. In *Genetic Algorithms and their Applications: Proceedings of the 2nd International Conference on Genetic Algorithms*, 1987.

[8]   J.M. Jenks. Non computer forecasts to use right now. *Business Marketing*, 68:82-84, 1983.

[9]   T. Kohonen. *Self-Organization and Assoiciative Memory*. Number 8. Springer-Verlag, Berlin, 1989.

[10]   N.H. Packard. A genetic learning algorithm for the analysis of complex data. *Complex Systems*, 4:543-572, 1990.

[11]   A. Stolcke. *Cluster Program Manual*. University of Colorado, 1992.

[12]   S.M. Weiss and C.A. Kulikowski. *Computer Systems that Learn: Classification and Prediction Methods from Statistics, Neural Nets, Machine Learning and Expert Systems*. Morgan Kaufmann Publishers Inc., San Mateo, California, 1991.

[13]   A. Scherer and G. Schlageter, A Multi Agent Approach for the Integration of Neural Networks and Expert Systems, *Intelligent Hybrid Systems* Eds. S. Goonatilake and S. Khebbal. John Wiley and Sons. In press.

[14]   P. Treleaven and S. Goonatilake, Intelligent Financial Technologies, Proceedings of Parallel Problem Solving from Nature, Applications in Statistics and Economics. EUROSTAT. 1992.

[15]   L. Blume and D. Easley, Evolution and Market Behaviour, Journal of Economic Theory, vol 58:9-40, 1992

[16]   P. Ormerod , J. C. Taylor and T. Walker. Neural Networks in Economics, *Money and financial markets*, Ed. Mark P. Taylor, pp341-353, Blackwall Ltd, Oxford, 1991.

[17]   Cheap and cheerful, The Economist, Feb. 27 1993

[18]   P.J. Kaufman, The New Commodity Trading Systems and Methods, John Wiley, 1987

[19]   W. Brock, J. Lakonishok, B. LeBaron, Simple Technical Trading Rules and the Stochastic Properties of Stock Returns, Technical report, The Santa Fe Institute, 1991

[20]   J. Lakonishok, A. Shleifer and R. Vishny, Contrarian Investment, Extrapolation and Risk, Technical report, University of Illinois, 1993

[21]   J. Kingdon, Neural Networks for Time Series Forecasting : Criteria for Performance with an Application in Gilt Futures Pricing, *Proceedings of the First International Workshopon Neural Networks in the Capital Markets.* 1993.

[22]   S. Goonatilake and S. Khebbal (Eds), Intelligent Hybrid Systems, John Wiley and Sons, 1995

[23]   E. Mamdani and S. Assilian. An Experiment in linguistic synthesis with a fuzzy logic controller. International Journal of Man-machine studies,7, 1975

[24]   H. Barenji, Fuzzy Logic Controllers. In R. Yager and L. Zadeh, editors, An Introduction to Fuzzy Logic Applications in Intelligent Systems. Kluwer Academic Publishers, 1992

Springer-Verlag
and the Environment

We at Springer-Verlag firmly believe that an international science publisher has a special obligation to the environment, and our corporate policies consistently reflect this conviction.

We also expect our business partners – paper mills, printers, packaging manufacturers, etc. – to commit themselves to using environmentally friendly materials and production processes.

The paper in this book is made from low- or no-chlorine pulp and is acid free, in conformance with international standards for paper permanency.

Lecture Notes in Artificial Intelligence (LNAI)

Lecture Notes in Computer Science